THE FOG OF DATA

Navigating Data to Derive Implications, Unlock Value, Get Buy-In, and Increase Transparency

JASON SCHENKER

THE FOG OF DATA

Navigating Data to Derive Implications, Unlock Value, Get Buy-In, and Increase Transparency

BY JASON SCHENKER

ISBN: 978-1-946197-27-6 *Paperback*
978-1-946197-21-4 *Ebook*

PRESTIGE
PROFESSIONAL PUBLISHING

*For people drowning
in data lakes.*

CONTENTS

CONTENTS

CONTENTS

War is the place of uncertainty; three-fourths of the things on which the business of war rests lay in a fog of a more or less large state of uncertainty.

It is therefore initially from there where a fine penetrating mind is required, with tact of judgement to feel out the truth.

Carl von Clausewitz — *On War.*
Translated by Jason Schenker

Der Krieg is das Gebiet der Ungewißheit; drei Vierteile derjenigen Dinge, worauf das Handeln im Kriege gebaut wird, liegen im Nebel einer mehr oder weniger großen Ungewißheit.

Her ist es also zuerst, wo ein feiner, durchdringender Verstand in Anspruch genommen wird, um mit dem Takte seines Urteils die Wahrheit herauszufühlen.

Carl von Clausewitz — *Vom Kriege.*

PREFACE

THE FOG OF DATA

This is a book about data. But it is also a book about challenges and conflict. Even the very title of this book is an allusion to these challenges and conflicts, as it references a quote about *Nebel des Krieges* — or the fog of war — from Carl von Clausewitz's book *Vom Kriege*.

But while this book was written to help people see more in an ever-thicker fog of never-ending data, it isn't a training manual on statistical analysis or how to use computer software to do statistical analysis or predictive analytics. There are many books for that, but this book is about conceptually framing the analysis you want to do, the way you talk about the analysis you do, and the challenges ahead posed by data on which we have become so dependent.

With the rise in data creation and data use, we have also seen some significant downside repercussions in the manipulation of social media and the use of fake news. And we see more challenges ahead with an arms race and weaponization of data.

I have been working with data, analyzing data, and creating data for over 15 years. I worked with data in grad school, investment banking, management consulting, and in my own firms, Prestige Economics and The Futurist Institute. It is a pleasure for me to share my knowledge and experience of data with you.

And I am quite excited that I get to do so in a futurist framework.

This book is primarily based on my experiences with data while running Prestige Economics and The Futurist Institute, including the work I have done in forecasting financial markets, the economy, company data, and industry data. This includes work I have done for Fortune 500 companies, large industry groups, governmental entities, and NGOs.

The main goal of *The Fog of Data* is to introduce you to and share with you what is most and least effective when working with data, doing forecasting, and communicating the implications of your analysis. Plus, it is my secondary goal to help you understand what risks will lie ahead as the pace of data creation accelerates with technological advances and the move to an IoT economy. In order to do this effectively, I openly share best practices that have worked for me, challenges I have experienced, and the pitfalls you need to avoid when working with data.

This book has been carefully crafted to provide digestible explanations of complex concepts. To meet these goals, I have structured a mix of explanations, anecdotes, pictures, and graphics to provide context and relevance to what can be an otherwise daunting set of data topics.

Acknowledgements

I want to acknowledge and thank all of the people who were involved in one way or another with the process of making this book come together. First, I want to thank **Nawfal Patel** and my other colleagues at The Futurist Institute and Prestige Economics who helped me bring this book to fruition.

Also, I want to thank **Kerry Ellis**, our cover designer, for bringing my ideas for the cover of this book to life. It was my intention that the cover would include imagery conveying mixed notions of data, technology, fog, and conflict. I am thrilled that we got there!

Finally, and most importantly, I want to thank my family for supporting me as I worked on this book. I am always most grateful for the support of my loving wife, **Ashley Schenker**, and to my parents, **Janet and Jeffrey Schenker**.

My family has supported me in countless ways over the years by providing emotional support and editorial feedback.

Every time I write a book, it's a crazy experience that spills over into my family life, so to them and to everyone else who helped me in this process, thank you!

Finally, thank you for buying this book.

I hope you enjoy *The Fog of Data*!

~ Jason Schenker

INTRODUCTION

NAVIGATING THE FOG OF DATA

The big idea of this book is that working with data and examining data is not easy. With all the data in the world available, we still have to have the right process, approach, communication strategy — and the right data in order to derive high-value implications. And the more data we create, the bigger the pitfalls can be.

A priori clarity and understanding is impossible when working with data you have not examined and analyzed. And these challenges are the wellspring of the goal of this book: To help you push into areas of data where you may lack clarity in order to minimize the data distractions you face, the analyses you need to redo, and the time you might otherwise waste.

Not all data is useful. And some data need to be ignored. But with ever-increasing amounts of data being created, this is likely to become a more critical challenge moving forward. And even while we may think we are working with all the data possible, there may be better data out there if we just ask — or if we just create it ourselves.

In order to tackle the most important factors in navigating the fog of data, I have divided this book into seven sections:

- **The Fog**
- **General Data Best Practices**
- **Data-Gathering Best Practices**
- **Data Analysis Best Practices**
- **Best Practices for Communicating Data**
- **Dangerous Data**
- **Pulling Everything Together**

In the first section of *The Fog of Data*, **The Fog**, I lay out some of the big issues with data — and why they are tough to see through. I discuss why I wrote this book and the importance of data analysis in Chapter 1. In Chapter 2, I discuss the increased creation of data that has accompanied new technologies and reduced costs of data storage and processing. Then, in Chapter 3, I discuss the dynamics of other data challenges.

In the second section, **General Data Best Practices**, I discuss some of the most important best practices when working with data. This includes a discussion of data process best practices as well as the importance of following an established data process in Chapter 4. Then, in Chapter 5, I discuss how to find strengths in the weaknesses of your business organization in order to do the best data analysis possible. In Chapter 6, I discuss three ingredients for personal success with data. In short, this section of the book should help you frame future analytical projects in a very tactical way. And this section should also have some tips to help you with your own long-term career planning.

The third section of this book is dedicated to **Data-Gathering Best Practices**. In Chapter 7, I discuss how to choose variables. Then, in Chapter 8, I discuss when to create data. This is followed by some recommendations on how to create data in Chapter 9. In Chapter 10, I write about the important of checking facts — and showing up. Part of good data collection is performing solid due diligence as well as personally gathering qualitative information.

The fourth section of *The Fog of Data* is titled **Data Analysis Best Practices**. In this section, I make the almost-radical assertion that not all data is important in Chapter 11. However shocking that may sound, I assure you it is true. Then, in Chapter 12, I discuss why it is important to ignore some data. In Chapter 13, I underscore how critical it is to get buy-in.

In the fifth section of this book, **Best Practices for Communicating Data**, I discuss something almost as important as gathering and analyzing data: communicating the implications. This is a really important — and often overlooked — subject. In Chapter 14, I discuss the importance of showing rather than telling when it comes to sharing data results. In Chapter 15, I discuss the value of telling stories to convey a message. But I also offer the critical caveat to not be weird when telling stories in Chapter 16. Then, in Chapter 17, I discuss how important it is to remember that this time is never different. And in Chapter 18, I discuss the most difficult subject for any forecaster or analyst: being wrong.

The sixth section of *The Fog of Data* is titled **Dangerous Data**; I discuss some of the nastier things in the data world, and I encourage you to be skeptical. In Chapter 19, I discuss how I have seen data misused in analytical work.

Then, I discuss the current state of fake news in Chapter 20 — and the potentially perilous and even more worrisome future of fake news in Chapter 21. Most people still do not know how social media works, how influencer marketing is structured, what paid social is, or how cheaply everything on the internet can be bought and paid for.

This section of the book draws heavily on talks I have given to the U.S. Department of Defense, the U.S. Joint Special Operations University, and NATO. Of course, I have also discussed these topics with corporate groups as well. But discussing the current state of fake news is something that still continues to shock executives and leaders — even at the highest levels. Pure democracy comes with costs, and while few countries are pure democracies (the United States most certainly is not), the press — the so-called Fourth Estate — has achieved a level of pure democracy. And without safeguards and understanding, the boon in fully democratized media and free speech may continue to present ever-greater challenges to maintaining a democratic and free society as a whole.

The final part of this section of the book — Chapter 22 — looks at the data, technological, and analytical arms races going on right now.

The most important of these from a political perspective is the analytical and technological arms race between the United States and China. The stakes are high — and rising. And both countries are being pulled into conflict in part by their individually self-interested pushes toward global technological dominance.

In this book's seventh and final section, **Pulling Everything Together**, I tie in themes throughout this book across topics to present some important actionable recommendations with corporate, industry, professional, individual, and global implications. I also share some other forward-looking futurist thoughts on the future of data as part of the book's conclusion.

In sum, the seven sections of this book should help you understand the most important dynamics when dealing with data. And they should help you manage the data you encounter professionally while also keeping you skeptical of the data you encounter in your personal life.

This book imparts practical strategies that you can implement immediately to make sure you, your team, and your company stay ahead of the competition by embracing technology — by successfully navigating *The Fog of Data*.

The Fog

CHAPTER 1

WHY I WROTE THIS BOOK

I live in a world of data. As an economist and a futurist, I look at data all day, every day.

If you don't work with data now, chances are you will sometime soon. After all, we are creating parabolically increasing volumes of data. And companies, governments, and individuals are increasingly digging into data for relationships, for business implications, and to make decisions.

I wrote this book to help people who analyze data, use data, and make decisions with data.

Today, that's pretty much everyone.

In this book, I wanted to share what I know and have experienced in a way that could help prepare others to avoid the mistakes I have made with data so that they can be at the top of their game more quickly. I started working with data professionally over 15 years ago. And I was working with it for several years before that in graduate school.

This book is a collection of insights that I have shared with numerous clients, industry groups, NGOs, and government entities in recent years.

In hindsight, I should have written this book years ago, but I have been busy building The Futurist Institute. In fact, a number of topics in this book are tied to The Futurist Institute's course on The Future of Data, which is part of our FLTA — Futurist and Long-Term Analyst — certification and training program.

Honestly, the main reason this book never got written is because I didn't carve out the time to write it. But now I have.

In the second half of 2018, I realized that The Futurist Institute had created so much research that it was time to codify the futurist implications of new and emerging technology dynamics in book format. This realization kicked off a publishing frenzy in September 2018, and I have since published one book per month. My goal is to do this for 12 months in a row. This is book No. 7.

And why am I the one to write this book?

My background as an economist and financial market forecaster is probably the most important qualification. In addition to my holding a Master's in Applied Economics, Bloomberg News has ranked me and my firm Prestige Economics one of the top financial forecasters in the world in over 40 categories since 2011. And Bloomberg even ranked me #1 in the world in 25 of those categories — many of them multiple times.

I am very proud to have forecasted many things well.

But I have made many errors with data over the years. And the biggest errors were often avoidable — or at least preventable.

There were countless times when I didn't have a good data process or I didn't follow a good data process. There were also times when I didn't have great data, but I didn't know how to work around those challenges. Worst of all, there have been many times when I have performed analysis that I was very happy with, but then I failed to communicate it well.

That's the worst! When you do some great work, but its unintelligible. And that, by the way, is completely on you. It's not your stakeholders' fault if you don't get buy-in, approval, or support for your analysis. That is your job!

You are the one who has to communicate things effectively.

It's important to know about best practices when doing data analytics, especially as data analysis becomes more common, more demanding, and more critical in the future.

In this book, we will also look at how your company, your profession, and you personally can better specify, collect, analyze, and share data. This is important because a lot of value can be unlocked from better data analysis.

But don't worry — we won't be doing crazy math in this book. And there are no statistical programming language drills.

It is my hope that by my sharing a mix of examples and anecdotes, you may be able to avoid some of the pitfalls into which I have fallen over the years. I hope that you can be better, faster.

That is really why I wrote this book: to help you avoid the mistakes I made to save yourself time — and headaches. And I have also shared a section to help you be more skeptical about data as well.

With that in mind, it's time for a data dump!

CHAPTER 2

TECHNOLOGY AND DATA CREATION

A Google executive speaking at a conference in Houston in October 2018 noted that the volume of data collected between 2016 and 2018 was greater than the amount of data created and collected in all previous human history.[1]

Businesses live and breathe data. They are expecting to find their next customers, cut costs, choose their next initiatives, and optimize their activities by analyzing data.

But the risk of computational data analysis paralysis is real if the data sets are too unwieldy for the computational powers at our disposal.

If the only solution to processing more data is *more* processors — rather than *better* processors — then the cost to analyze data is going to be an issue that increases with the volume of data. In other words, the cost to buy more processors to get the job done could also go parabolic. And that is becoming a real risk.

The End of Moore's Law

Part of the reason for the massive shift in increase of data has been a massive decline in the cost of data processing and data storage hardware. But that may be coming to end — especially for processing power costs.

This is because computational powers are facing some potential limitations. In fact, many people working in technology will speak about a big near-term risk, and that's the limit of something called Moore's Law, which is named after Intel founder Gordon Moore. It is the notion that computational powers can be doubled while the costs are halved.[2]

This has made increasing computer processing power technology both more powerful and cheaper. A graphical depiction of the development of computer processing power along the lines of Moore's Law is depicted in Figure 2-1.

But Moore's Law is breaking down.[3] The computational increases with lower costs just aren't happening anymore.

In truth, the only solution now is "more processors" versus better processors. And this is a really big problem as the amount of data being created accelerates. After all, if you want to be able to analyze data you collect, you need processors that can handle the amount of data you are collecting.

This is going to become more of a challenge as the amount of data we collect and store continues to rise parabolically, especially in an *internet of things* — or IoT — world.

Currently, we have an internet of computers, tablets, and cell phones. Those are the primary devices that have internet connectivity and through which data can be collected and tasks can be enacted.

But the drop in sensor costs and a desire for increased consumer convenience is going to usher in an era called the *internet of things*, when many *things* are connected to the internet, not just your computer, tablet, or phone. This could include your car, refrigerator, and office supply closet.

Each of these devices will interact, take actions, and generate data. When all devices are connected to the internet, the amount of additional, minable, analyzable data will be mind blowing.

Figure 2-1: Moore's Law[4]

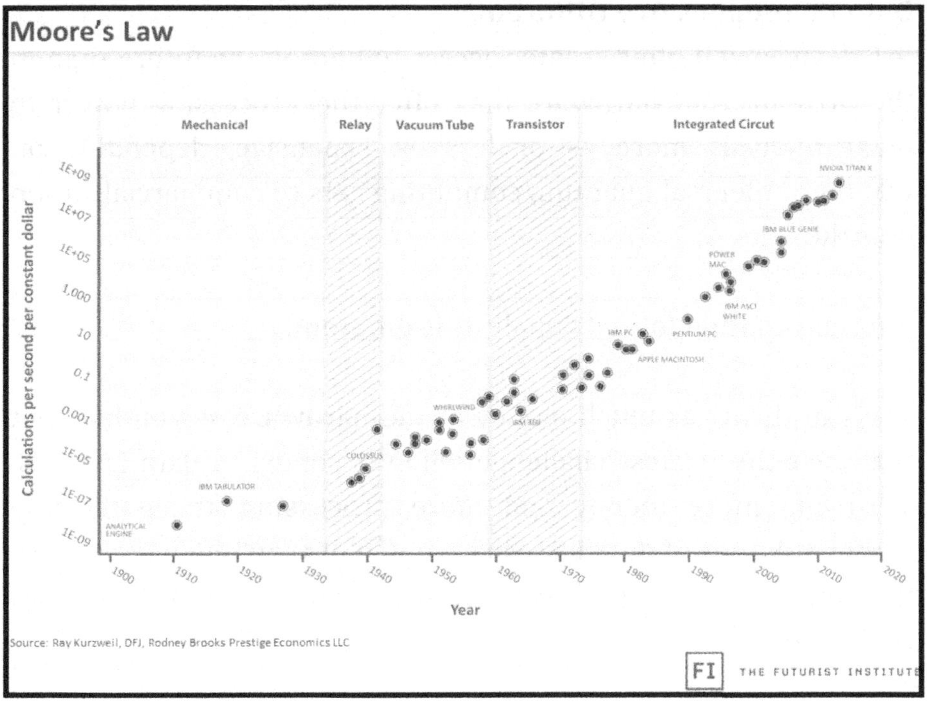

Current levels of existing data collection and analysis will likely pale in comparison to future IoT data.

Now, when it comes to processing power, without a step change from technology like quantum computing, the only strategy available would be buying more processors rather than making significantly cheaper or better processors. This is talked about in technology circles as a kind of "brute force" solution.[5] Technology experts, scientists, and futurists use this term because buying more processors isn't inventive.

It's a problem solved by money rather than scientific innovation. It's just throwing more processors at the problem rather than pushing for a step change in computational processing power.

Data Challenges Are Different

IoT devices will offer greater access to data in the future than in the present. And hardware may offer the processing power to make analysis more — or less — expensive, depending on whether universal quantum computing gets to commercialization quickly.

But the situation for data analysis is different.

You can throw as much money at the hardware as you like, but that isn't the biggest challenge when it comes to data analysis. Throwing more money and more processing at significantly higher volumes of data just isn't enough. It won't help make sure the analysis is good. It won't solve the actual biggest data challenges.

CHAPTER 3

DATA CHALLENGES

Despite my discussion in the previous chapter about technological opportunities to generate more data, as well as the computational limits of processing power used in analysis, the biggest data challenges, problems, and opportunities ahead lie with people. They lie with you.

Some of the most critical issues and data challenges I have seen in my career include the following:

- **Data analysis does not follow a rigorous and consistent process.**
- **Data gathering is not properly specified from the outset.**
- **Statistical analysis, predictive analytics, and forecasts fail to perform, because the wrong questions were asked.**
- **Communication about data analysis is ineffective.**
- **People don't look at analysis with enough skepticism.**

As you can see, there are a lot of challenges, but the good news is that these challenges can be proactively addressed.

In the coming chapters of this book, I will discuss these challenges in more detail. And I will provide proactive strategies and solutions for how you can prevent them.

But before we get to addressing the biggest data challenges, let's consider the five steps in a data analysis project.

When most people talk about data challenges, it is often wrongly assumed that the biggest challenges will occur in the analysis part of the project. And making sure the most important data gets collected and analyzed is a major focus of the next chapter. But I assure you, the biggest challenges can and often do occur elsewhere.

Where the Wild Data Challenges Are

The two biggest challenges with data analysis are the risks that data analysis does not follow a rigorous and consistent process and that communication about data analysis is ineffective.

In the next chapter of the book, I outline the importance of a process, so I won't belabor the point here.

But as to the topic of communication, I cannot underscore how big a problem this really is. Economists, analysts, and statisticians are often viewed as a geeky bunch talking about detailed stats to audiences that may not be as versed in statistics. In my career, I have tried hard to avoid discussing the detailed attributes of various statistical analyses and forecasts because the group of individuals who are usually most interested is at the executive level.

And I have, unfortunately, seen data analysts and forecasters discussing detailed statistical attributes of analyses with an audience that involved executives, most of whom had not likely taken a stats course in 30 years. That kind of audience wants the high-level framework, the inputs, and implications — not a discussion of kurtosis and adjusted R-squared terms.

When you do that, it can be extremely counterproductive — and it can greatly diminish your ability to get buy-in.

Another example of major data challenge tied to communication is the awful tendency among technology experts as well as futurists to use buzzwords. Those of us who are active in the technology world often refer to this as buzzword bingo.

An example of this is the way you can hear people often use the terms *artificial intelligence, machine learning, predictive analytics,* and *big data.* We are still quite a way from true artificial intelligence, which means for now, most of the data analysis to derive valuable, actionable, and potentially profitable implications comes from data machine learning or predictive analytics. Essentially, they are the same thing. And it is something that most people simply called statistics up until just a few years ago.

When it comes to communicating the implications of data analysis, the best thing you can do is get to the point quickly and avoid being an aardvark. I'll discuss this in more detail in Chapter 16, which is titled "Don't Be Weird."

General Data Best Practices

CHAPTER 4

DATA PROCESS BEST PRACTICES

The most important thing to remember when analyzing data is to do what is necessary, required, and useful. This is how I approach sharing market analysis with my retainer clients, and it's how I share modeling and forecast results when we do special projects.

And it all begins by asking the right questions.

Asking the Right Question

In the movie *The Hitchhiker's Guide to the Galaxy*, some of the main characters go to an all-knowing computer named Deep Thought and ask it a big question. These characters are seeking "the answer to life, the universe, everything."[1]

After a period of time, they return to the all-knowing computer for the answer, which is given as "42."

Of course, the characters didn't ask a good question. And without a good question, you will not get the answer you want either.

Without the right question, all the data and computational power in the world would be worthless. And while the right question is critical, making sure the data is sound and that you follow a proper data process is also critical.

Following a Data Process

Two things make following a proper data process more critical than ever before.

It's important to remember there is a human side and a tech side to every data project — and it is important to properly conduct analysis from a technical and a process standpoint.

The rate of data creation and collection is accelerating, and 90 percent of the world's data has been created in the last two years. This means that scientists, engineers, and businesses have an incredible opportunity to leverage that data to their advantage. However, they can only harness the power of that data to yield real, valuable, and actionable insights if they follow a process.

And mistakes of bad data and bad process management illustrate why it has become increasingly important to correctly manage the entire data collection and analysis process so that everything flows smoothly. It is important so you are sure to get valid results that support — or refute — your hypotheses.

Making sure to follow a solid data process consistently is key to achieving the best analytical outcomes. And it will save you time and effort if you measure twice and cut once when working with data.

Data Process Best Practices

Following a proper data process means that you need to follow a series of steps to get the insights you want. This means you can't just ask the right question. You also need to collect the right data, make sure the data is clean, and use the right data. It's only then that you can actually analyze the data. And after the question is right, the data is right, and the analysis is done, you still need to test and retest the results.

I've been analyzing different economic, financial, and statistical data for over 15 years, and I recommend consistently using a seven-step data process, excluding the presentation and use of the data:

1. **Formulate the right question**
2. **Scope the project**
3. **Collect the data**
4. **Clean the data**
5. **Analyze the data**
6. **Test the results**
7. **Retest the results with new and future data**

Let's go over this framework — a framework that will only become more valuable when the power of quantum computing is at our disposal.

Keep in mind that following this process in order is equally as important as following the steps. If you go out of order, the value will be lost.

Planning to Formulate the Right Question

Albert Einstein said that if you have an hour to solve a problem, spend the first 55 minutes formulating the question and five minutes getting to the answer. With quantum computing, that ratio may be even more skewed, where you may wish to spend 59 minutes formulating the question and just one minute getting to the answer. But whatever the ratio, the implication is clear: You have to start with the end in mind. You have to start with the right question.

And whatever amount of time you have to formulate a question, one thing is certain: The more time you have to formulate the right question, the more value you will derive from data. This is likely to be even more true for quantum.

Of course, coming up with the right question is actually sometimes the most difficult part of an analysis project!

One thing classical computers as well as quantum computers — whether true quantum, emulated, or photonic — will be unable to do for a very, very long time to come is ask the question you want answered.

You need to come up with the right question yourself.

Of course, you'll need to make sure that the question is specific enough to be answered. And you'll need to be sure that whatever data you decide to collect helps to answer the question.

I've found that it helps to think about a potential question from all angles as part of a brainstorming process. This will also help you come up with preliminary ideas for what data you think would help you answer that question.

For example, formulating forecasts for long-term oil prices, you would need to consider future supply from conventional and shale oil wells, the future demand from rising emerging market wealth, potential future demand abatement from electric vehicle use, and future financialization across financial markets and how that behavior could influence oil prices.

Scope the Project

Of course, before you decide exactly which data you need to collect, you have to scope your data project.

For example, you may not want to dig deeply into international data when your project or client is primarily concerned with the United States or a particular state. This is like the aphorism I've already alluded to earlier in this chapter, that you should measure twice, so you only need to cut once.

So, while you may be looking at oil prices, you may not want to be trying to build forecast models for every grade of crude oil or every petroleum product at once. I mean, maybe this is something that a quantum computer could help you with, but for the moment, if you need the computation done in a reasonable amount of time using a classical computer, you may want to make sure to put some limitations of scope around your analysis.

You should be aware that although scoping a project may sound straightforward, there can be major challenges in working with stakeholders.

One of the biggest challenges in scoping a project and working with stakeholders is that there can be a lack of clarity. It's not uncommon to work with stakeholders who may want something that doesn't exist — or they may want something that they do not fully understand. This is not an uncommon situation if you are working with executives who are outcome-oriented.

If you come across these kinds of scoping issues, it is up to you to navigate this kind of professionally perilous fog quite carefully. It's important to remember that not everyone may have your specific knowledge of data, statistics, or analytical tools. But it is also important to keep in mind that those are your stakeholders — and that their knowledge and experience likely dwarfs yours in other areas.

My top piece of advice about this is to be nice. I know it's a weird thing to say when talking about data, but be nice to your stakeholders — especially if they don't fully understand the details.

There are other risks with stakeholder interactions as well, and these can include confirmation bias, anchoring bias, and other biases that can become problematic later in the process.

In the case of confirmation bias, they may be looking for analysis to confirm beliefs they held before the analysis even begins.

In the case of anchoring bias, you may find that your stakeholders are fixated on a certain expectation — or even a certain very specific growth figure, Fed funds rate expectation, or timing of the next recession. You may have provided these numbers — or they could have come from somewhere else, unbeknownst to you.

There are other biases that can come up in data analysis stakeholder interactions as well, and the best thing you can do is address the issues one by one. You may feel like you are beating a drum that no one hears, but the best way to overcome bias is to be consistent and reliable in your messaging.

Collect the Data

After you have identified the right questions to ask and you have scoped the analysis you want to do, the next step is to collect the actual data.

This means that you also need to think about where your data is coming from. If you are using data that you specifically had created to answer your questions within the scope of the project, then that is going to be a huge advantage over trying to make data that has already been collected for another purpose fit perfectly.

If you are using internal data, then you need to make sure that the data is appropriate. If you are using outside sources, you need to make sure they have valid data that you are allowed to use. External data might include government or financial market data.

Although you may have access to a veritable treasure trove of data when you get to this point, you always want to go back and think about the question you are trying to answer with the data. If the data you have can't answer the question, then you might need to work on creating data that you can collect to answer your question.

You should be aware that there can be many different issues that crop up during the process of data collection.

You can have issues with government data that stem from a slow collection and reporting process. Or you may find, as I recently did, that a government shutdown delays the publication of critical data.

Even worse than occasionally unreliable government data, however, is the chance that there may not be any official data related to what you are trying to analyze. In this case, you may need to create data. This is a topic I discuss in greater detail in Chapters 8 and 9.

You may also find that there is a suspect source of data or that there may be inconsistency of data collection, cohort, or commitment. These are big challenges that should reasonably hinder your ability to trust certain data blindly.

Similarly, you may find that the data you are working with is comprised of incomparable data sets, which could lead to an apples-to-oranges comparison. You need to be very careful about this kind of data problem.

After all, if the data sets don't match up and you do analyses, the implications could be misleading, outright wrong, and straight-up worthless.

Checking that the data you are working with is appropriate, reliable, and trustworthy is a necessary precondition of the next step in the data analysis process: cleaning your data.

Clean the Data

I've already alluded to making sure you're using the correct and appropriate data to answer your questions. But you also need to make sure that the data you plan to use is also "clean" — or consistently and accurately recorded.

Data that isn't clean can taint the results of any analysis you might wish to conduct. This is why cleaning data is a critical part of any data analysis process that you don't want to skip.

This means that your data needs to be consistent, uses the same units of measure, is for the same time period, and is formatted correctly.

Some questions you might ask as you perform due diligence on your input data might include the following:

- Is everything in the right currency?
- Do the data show the right units of measure?
- Are the columns and rows of data all in alignment?
- Are the attributes of the individuals in the panel data consistently labeled?

If there are inconsistencies in your data, you may need to make adjustments to the data, eliminate some data from your analysis, or select new data you could use or create to perform the analysis required.

Cleaning your data is critical because it will help to make sure you don't run into any issues when conducting the analysis. It also improves the usefulness of any analytical implications, and it can prevent your need to redo analysis from scratch if data problems are discovered later.

It can cause a huge headache if someone skips ahead to analysis without cleaning the data to be compatible with the technical tools being used.

Misaligned columns can yield faulty results when run through a statistical software package. And they could lead to misinformed corporate decisions, investments, or strategies.

Especially in a world of cloud computing, where you may be paying for the ability to complete an expensive or high-value analysis, you don't want to waste your time and on-site or remote processing power by putting data that isn't clean through the process.

That's a waste!

Have you ever heard the phrase "Garbage in, garbage out"? This is what happens if your data isn't clean. Your results might be garbage.

Only after your data is clean can you move on to actually analyzing the data. And if your data is not clean, then you may need to recollect the data. After all, you want to make sure that your analysis yields something fruitful, which is highly unlikely, if the data is all messed up. Such data that isn't clean is also referred to as "dirty" data.

Analyze the Data

This is the part of the process where a classical or quantum computer actually needs to bring its A-game. An increase in computation power and greater storage capacity has augmented the use of data and analytics. And continued investing is further pushing the boundaries. But while analysis is getting easier, staying on track could become tougher for the person acting as Sherpa in the analytical process.

Perhaps not surprisingly, the actual analysis is the part of data analysis projects that many people are focused on. And it's where quantum computing offers the potential as a future technological step change.

But the potential for current classical computing — or future quantum computing — to bring valuable data analysis will still fundamentally depend on how appropriate, good, and clean the related data is. And it will be highly dependent on solid data process management.

Honestly, analyzing data is currently one of the easiest parts of a data analysis project.

Hey, cleaning data isn't easy.

And collecting data you don't have can take a long time, especially if you need time series data across multiple time periods and cannot collect historical data or reconstruct the data you need ex post facto.

One recommendation I would make is that no matter what kind of analysis you're doing, is that you revisit the question you are trying to answer. Sometimes data can get you off track, so it's important to focus your efforts on the task at hand.

Similarly, big challenges can occur in the analysis part of a project due to issues unrelated to statistics. Sometimes there are other priorities for analysts. And there can be managerial pressures for analyses to play into existing biases.

There are also challenges that can stem from issues with time constraints that can reduce analyst effectiveness. Analyst role changes can also reduce model efficacy. Plus, and more disconcertingly, there can be moral hazard risks that stem from conflicts of interest, and there are also dishonest analysts out there.

But assuming that you clear the minefield of these human-oriented risks of data analysis, the analysis still isn't complete after you have an answer.

You need to test your results — repeatedly.

Test the Results

After you complete your analysis, it's important to test your results. This is also an important part of the process. After all, you need to determine if the analysis is right. To do this, go back and reconfirm the other steps you conducted before the analysis.

No matter how much data is created or used, the process is likely to remain the same. It's just the type of testing, the frequency of testing, and the complexity of testing that will change with advances in technology like quantum computing.

But the process will be the same.

Retest the Results with New Future Data

I do have one last bit of bad news in case you thought you were done with your analysis and that you could encase your models in amber and allow them to exist statically for hundreds of years or more.

You need to retest the model to see if it works with new and future data. Data relationships can change over time, and when you get new additional data — or when more data is created in the future — any models or analysis will need to retested. This is especially true for dynamic data sets, where a multitude of factors are likely influencing the data that is being created.

Consider some dynamic data, like real-time optimization of massive e-commerce supply chains or the information received from self-driving cars.

Retesting and reshaping analysis and the way data is analyzed, as well as the implications thereof, could actually become more efficient or even possible with quantum computing processing power. This is because some of the data sets with the most information available may currently be intractable to perform in real time or in a reasonable amount of time to be considered useful.

This is another potential benefit of quantum.

Retesting will have to be frequent to allow for adaptations. This means that choosing the right data and cleaning the data before its use will become even more critical parts of the process — especially for instantaneous, iterative data analysis that may have real-time implications in the real world for large systems like supply chains, transportation networks, and medical treatments.

The fundamentals of the process will remain the fundamentals no matter what.

Other Issues

Following a process to perform analysis, as I have laid out in this chapter, is the most critical piece of a successful analysis project. But it is not the only piece. These are other critical elements required to actually be effective. After all, you still need to communicate the implications of your analysis. And if the communication piece is ineffective, then all of your solid analysis could prove worthless. The effective presentation of data and helping stakeholders take advantage of your implications is discussed at length in Chapter 14.

Summary of Good Data Practices

Here is a summary of the best practices of the data process I have outlined in this chapter:

- **Data without useful questions to answer is useless.**
- **Good questions without the right data cannot be answered.**
- **Good questions with the wrong data may be answered wrongly.**
- **Good questions and good data with poor analysis is worthless.**
- **What constitutes "good" for data analysis must be continually tested and retested.**

It's also important to remember that when conducting data analysis, you have to follow the process in order. It makes no sense to analyze data you haven't cleaned. And it doesn't make sense to clean data you haven't identified as in scope for the prioritized analysis you are looking to perform.

As you can see, there are technological limitations, physical science limitations, and human limitations to the current and future efficacy of classical and quantum computing.

While many people in the field of data are focused on pushing the limits of computational processing power, the most important limitations of computing will be the same for a long time to come: making sure the humans in the process structure data analysis properly.

CHAPTER 5

FIND STRENGTH IN WEAKNESS

One of the most important things I can share about my own experience with doing data analysis is that it is important to level the strengths you have. Maybe you have access to a lot of data, very powerful software, or a big team that is highly competent. Whatever it is, however, you need to make sure to level those strengths to the best of your abilities.

This is also something that you can do with your weaknesses.

I run a small research firm, Prestige Economics, which means that there are some pretty big challenges. For example, I do all of the forecasting and writing of our research reports. Of course, I don't do all of the work. I have some amazing people who support me, but when it comes to the actual forecasting, analysis, and writing it's just me. This presents a downside weakness known as a key-man issue. It means that the work depends on me as a key element. But I have never missed a monthly report deadline in almost 10 years, and I don't ever expect to.

There are also two other big weaknesses for my business as well.

One of those weaknesses is that my independent firm does not have the distribution or the platform of a large investment bank, bulge bracket consulting firm, or large research entity.

The third and final weakness of my entity is that we have a smaller number of clients. As a small firm, you can serve only so many clients, and that is a limitation of sorts.

The reason I want to share these weaknesses is that in many ways, when it comes to data, information, and analysis, these weaknesses provide strengths. And I find it is really important to consider the flip side of your weaknesses regardless of your situation.

So, while my firm may have a key-man issue, lack a large platform, and have a relatively small number of clients, these are actually some of our greatest strengths.

The inverse corollary of a key-man issue is that our research can be written and prepared rapidly, and a consistency of storyline and analysis is assured. Because my entity is small, we are agile in our deliverables and timing. And because we have a small number of clients, we are able to give our clients more attention. And they have direct access to me personally. Clients of larger research or consulting firms seldom have access to the actual thought leader within the organization.

This is a common phenomenon when working with data and producing research of all kinds. Weaknesses can offer insight into strengths. But strengths can also betray weaknesses.

Not too long ago, I gave a presentation to a large governmental organization that had all the strengths my firm lacked: It had a large organization, a very large budget, and many stakeholders. Those were its strengths.

Interestingly, its strengths were also a source of weakness.

After all, a large organization can slow down analysis because of a need to coalesce and reconcile data and analysis from different individuals and groups. And there is a risk that some analysis may be perpetually siloed and never reconciled. Plus, a larger organization with a big budget may find that there is reduced budgetary control. So, even though a large organization may have massive resources, it may be slower to allocate it — if at all.

I give these examples openly so that you can consider the strengths of your organization as it approaches data and research. You may have access to massive amounts of data. That is a big strength. But you also need to manage the weakness that comes with massive data. There will be more data to collect and clean — and you will need to coordinate between departments, which can make the process slower. If you can manage those downside weaknesses, it can help you unlock the strengths you have.

Similarly, if you have only a small data set, but it is clean and comes consistently from a trusted source, you will not have as much of an issue with collecting and cleaning your data, but your dataset will logically not be as robust. Trying to find ways to manage that weakness will be critical.

In the end, analysts seldom have all the data, support, or infrastructure they want. But that's just the reality of data.

There is a saying about warfare that seems somewhat apropos: You don't always go to battle with the army you want; you go with the army you have.

The same is true of data. You don't always have the data, infrastructure, or support you want. But you have what you have. And you have to find the advantages and benefits that are inherent to the conditions in which you are operating.

For me, that means being agile, consistent, and having close working relationships with my clients. When I worked in a larger entities in my career, like McKinsey and Wachovia (before it became part of Wells Fargo), I had massive resources that allowed me more leverage but that also often involved a slower process of publication, research, and client communication due to layers of compliance, control, and management.

There is an ancient Greek saying about people's lives as a hearth. On the hearth there are two vases. One is for the good in life. And one is for evil. Everyone has two choices: Either a mix of good and evil — or only evil. No one gets only good.

And so it is with data analysis. There is no such thing as perfect data, perfect support, or perfect communication. Find the mix of strengths and weaknesses and leverage the strengths of your business to find the most upside — and navigate the fog of data.

CHAPTER 6

THREE INGREDIENTS FOR PERSONAL SUCCESS WITH DATA

In addition to following a good data process and playing to the strengths of your organization, you also need to be proactive in making sure that you experience personal success with data analysis.

The three most important ingredients for success are as follows:

- **Perpetual learning**
- **Significant external professional activities**
- **Writing books and articles**

Perpetual Learning

The most important personal ingredient of success with data and analysis is perpetual learning. You need to be familiar with statistics and analytics in order to perform data analysis, but as you are asking the right question and scoping the analysis you want to do, you may need to draw on a number of different intellectual disciplines. After all, if you are expected to yield innovative insights with your analysis, you have to have the knowledge base to present something informed and innovative.

This is something I think about all the time.

Because I analyze financial markets in regular reports, people often assume I am in New York. But I'm not. In a modest callback to the last chapter, this may be seen by some as a weakness, but I find it is a strength. After all, many analysts who look at financial market data are looking at the same data as everyone else. They are sitting at a Bloomberg terminal in Midtown Manhattan. And if they leave work and interact with other people in the same milieu, looking at the same data, who live in the same place, this actually diminishes the potential for interactions that broaden the economic financial context in which the data that streams across a computer screen exists.

Being based in Austin forces me to make choices about traveling to visit clients and do research. I have to get out there in order to learn what's going on beyond my screen. And I would argue that being removed from New York reduces groupthink risk, but it also means I have to do more individual legwork to test theories on markets and the economy.

I will discuss this concept further in Chapter 10. But for now, the big takeaway is that learning isn't just something you do from reading books. It is also something that comes from talking to people, going to conferences, and meeting with decision makers.

I cannot stress enough how important it is to budget time and money every year to learning new things. This is especially true if those things tangentially relate to your work and data analysis.

By learning new things and exposing yourself to new ideas on a regular basis, you can be better prepared for data that is unexpected. You can be better prepared to bring new ideas about how to analyze data, scope analysis projects, and set priorities on what data should be collected.

Very few people working with data are constantly asked to look at the same data and do the same thing to get the same results. After all, if the analysis were that easy and clean, you could just write a script for it so it could be automated.

And that just isn't easy to do with most analysis.

In most cases, you are being asked a new question, with data that may or may not provide you with insights. And you may have to come up with an unconventional approach to the problems you are given.

Although statisticians, econometricians, data analysts, and other quants are often maligned as being very rigid due to their analytical professions, the truth is that data analysis is sometimes a very creative process.

There are constantly new challenges, new goals, and new problems to attack. This is why it is important to be perpetually learning and why it's important to block out time in your annual calendar for research trips. Because you will be expected to provide innovative insights. And those don't come from rote memorization and perpetual repetition.

One way to build your analytical data chops is to pursue formal education in the form of degrees, complete certification programs, or take online courses. I think that a mix of formal and informal education is best, because the goal-oriented and deadline-oriented nature of degrees and certifications forces you to carve out learning time. But informal education opportunities also present the opportunity to resolve immediate-term tactical knowledge needs.

Whatever path you choose, learning new skills and taking new courses is likely to prove beneficial to your work — and it's likely to be financially rewarding.

Significant External Professional Activities

The second area where you can help make a personal growth commitment and significant contribution to your career in order to strengthen your data analysis skills is in building up your external professional activities and network. This echoes somewhat my notion of perpetual learning, but it places it in an activity-based framework.

And you don't have to do it just within your company or industry.

I have been involved with a number of professional organizations over the years, and I would note that company, industry, and trade association meetings and conferences offer you the opportunity to interact with professionals who may be facing challenges similar to the ones you face. But cross-industry functional conferences are likely to yield the most meaningful and innovative insights.

For example, if you are a data scientist or econometrician at a retail firm, you would probably gain more from an activity that is external to your company and your industry but which is more directly tied to your data analytics function. So, rather than getting more involved in retail industry conferences or meetings, you may find that interacting with data scientists and economists across industries — as a functional group — may be more beneficial.

During the course of any year, I attend events with economists as well as events with other CEOs. This allows me to gain insights from the challenges and innovations that are being deployed in other situations on a noncompetitive basis in order to do better analysis — and be a better leader.

Writing Books and Articles

The final ingredient to making yourself a successful analyst is sharing your insights. One of the best ways to learn is by teaching others, and writing books and articles is a great way to generate opportunities to teach others. But it is also a great way to learn from others — and it provides the opportunity for interactions with other analysts at conferences, in your industry, and virtually.

Sharing what you have learned helps others, and it can prove beneficial for yourself as well.

In sum, these three strategies of perpetually learning, engaging in significant external professional activities, and writing books and articles are important ingredients to achieving data analysis best practices. They are ways to build your skills and hone your craft.

Data-Gathering Best Practices

CHAPTER 7

HOW TO CHOOSE VARIABLES

When doing an analysis, choosing variables is a critical part of the process. In my experience, this is one of the most difficult parts of the analytical process.

Sure, you need to follow a process, and you need to lever the strengths of your business operations. But this is where the road meets the pavement. Choosing variables is when you need to tap into your perpetual learning process in order to create the necessary and sufficient conditions for innovative thought.

Or at least for directionally correct thought.

There are four main priorities when it comes to considering variables, at least in my experience:

- **Inputs that make sense**
- **Leading variables**
- **Directional consistency**
- **Model fit**

Inputs That Make Sense

The first prerequisite of data analysis is that it needs to make sense.

I often invoke the notion of Occam's Razor when making this point in discussions. Named after a mediaeval monk, Occam's Razor is the notion that the simplest answer is often the right one. Although those that favor deductive logic may take umbrage with this assertion, it is a notion that should not be overlooked when doing data analysis.

After all, at the end of the day, the data relationships that you test and ultimately identify as statistically significant — and hopefully predictive in nature — should make sense to a certain degree to anyone with whom you may have the occasion to discuss your model and data.

This doesn't mean that your mom needs to understand your logic, but that is a pretty good barometer that I live by.

In any case, it does mean that people in other departments or other parts of your organization who have little knowledge or understanding of statistics should be able to understand the logic behind the models you constructed.

In essence, this just means that your models and analysis should make sense to an outside observer.

Let me give you a few examples.

First, let's talk about commodity prices. When I look at creating forecast models for commodity prices, one of the most important elements — in truth, the most important element — is usually Chinese manufacturing activity. In statistical models I have used over the years, this is a critical input for commodities like aluminum.

Why does this work?

Because China consumes a majority of the world's aluminum. And as it turns out, commodity demand is more sensitive to growth dynamics than supply. That means that supply will typically be slower to respond to demand shocks — either on the upside or the downside.

This also makes sense because it is very costly to open a new aluminum production facility. It is also very costly to shut one down as it will take immense financial and business will to reopen it should you need it.

This means that supply may continue to run high even as demand falls, further exacerbating downside price risks. Conversely, supply may hold back even when demand runs hot because the expense of opening a new facility needs to be justified by forecast models that will likely only show profitability at relatively high levels.

So, if China consumes a majority of the aluminum in the world and demand drives prices more than supply, watching Chinese manufacturing figures is logically critical. It just makes sense.

Of course, any data analysis may not be as simplistic. But you should be able to explain it with similar ease. If you are trying to do predictive analysis on when a facility may have high demand, you should try to find something that logically matches with high demand. That could be anything from seasonal factors — like online holiday shopping — to broader drivers of demand, like strong GDP, low unemployment, a weak dollar (for U.S.-based companies), the stock market, or low interest rates.

Leading Variables

Assuming that you have variables in your models that made sense, it is important to try and make sure that you have some leading component to your models.

For example, let's say that you want to predict the monthly change in U.S. retail sales jobs. After doing some analysis, you find that monthly U.S. retail sales data is correlated with monthly retail sales jobs. So, if retail sales are strong in March, then retail jobs may also rise in March. But there is a key problem with this analysis: The March retail sales report comes out after the March jobs report.

In other words, you will have discovered a correlation, but the retail sales report isn't a leading indicator — it does not offer future predictive value. In fact, it is a lagging indicator. This would be another case where you need to go back to the drawing board.

In short, you need to make sure whatever you are predicting has data that is available beforehand.

If you go back to your modeling and find that retail sales jobs data actually lags one month behind retail sales, you could use that to make a prediction. Similarly, you may find that a moving average or moving total of retail sales is predictive of retail jobs data. Or you might find something else entirely different that leads retail jobs data, like consumer confidence.

The most important thing to remember is that whatever inputs you are using in your models, they have to make sense, and they have to lead whatever it is you are trying to predict.

In fact, you may find that when predicting trends and dynamics in economic data, financial markets, or corporate data sets, you need more than one variable.

Directional Consistency

If you are trying to analyze the dynamics of your data, you may often find yourself looking at a mélange of factors rather than just one factor. This mix of factors is usually analyzed in what is called a multivariate regression analysis. In other words, there are multiple variables. For reference, analysis using only one variable is called a univariate analysis.

When you look at multiple elements, it is important that they have directionally correct impacts. By that, I mean both individually and in concert. The multivariate regression and analysis variables should behave in a way that is logical.

For example, if you are trying to forecast the demand for your exports of auto parts to Europe, you would likely want to consider the strength of the euro. A stronger euro means that your dollar-denominated parts would be cheaper than if the euro was weak. In other words, more parts can be bought with a strong euro. In light of this, you would expect a univariate relationship of your parts sales to be positively correlated with the strength of the euro.

Let's say you also find another factor is important for European demand, like a low import tariff rate. This would be another factor that would be inversely correlated with demand from Europe. In other words, the lower the import tariff, the higher the demand to import your goods would be.

Tariffs and exchange rates are usually going to be uncorrelated. This means that if we do a multivariate analysis of your parts sales into Europe, we are looking at performing a regression where sales are a function of the euro and the tariff rate, such that

Sales = *f* {Euro, Tariff Rate}

Now, we know in the univariate analyses that sales into Europe are positively correlated with the euro. And we know that sales into Europe are inversely correlated with the tariff rate. In models I build that have uncorrelated factors, it is normal to expect that the directional impact of each of the variables should be the same. In other words, the strength of the euro should have a positive impact on exports in both a univariate and multivariate analysis. Conversely, the level of tariffs should have a negative impact on exports in both a univariate and multivariate analysis.

If a multivariate analysis of these two uncorrelated factors shows that a strong euro is bad for imports and high tariffs are good for imports, there is something wrong with the model. And you may need to look for a counterintuitive factor, or, more likely, you need to consider adding in additional variables or scrapping the model in place. The format of the variable inputs in a model need to make sense, whether alone or in a combination with other variables.

One of the key words I have used here multiple times is the word *correlation*, and multicollinearity can be a problem when doing statistical analysis. This is a deeper and more advanced subject for those who model data. But the essence of this issue is that data may not be uncorrelated. And if there is a correlation, it can make modeling trickier — and it needs to be addressed and controlled for.

Model Fit

Assuming that you have nailed down the other three data prerequisites, your final goal is to make sure that your data provides a model with a good fit.

For some analysts this may mean a certain margin of error. For those with a more statistical bent, this may mean a certain adjusted R-squared, T-test value, and root mean-squared error.

No matter how one would describe what they are looking for in a model fit, the analyst is essentially looking for a model that is predictive and that can be helpful for predicting whatever the objective of the analysis is.

The most important factor to ensure the model has the potential to be a good fit is a robust data set.

The more data you have, the more likely it is that that the data will prove predictive in a wide variety of situations. In some cases, you may have too limited a set of data, and this can call what analysts refer to as an overly tight fit in the model. This means that it can predict the past well, but you may find it not very useful for predicting the future.

That makes sense, right?

If you have a more limited data set, the ability to predict effectively will be hampered by a lack of information and experience captured in the historical data set of inputs for the models.

Of course, you need to follow the process for good data analysis, and you should also have the other priorities locked down — inputs that make sense, leading variables, and directional consistency. You may need to keep trying different combinations to find a set of variables that follows the right process and matches the priorities.

But in the event that you do not have data that produces the information you want for the model to be a good fit, you may need to consider the potential need to create your own data.

I discuss this notion in the following two chapters.

Further Reading

I have taken a number of liberties in discussing statistical analyses in this chapter. And I do not want to give you the impression that everything I have touched on is what you need to know. In truth, I did an entire master's degree on this subject and needed to learn even more after that was complete in order to effectively forecast and build predictive models.

My two strongest recommendations for further reading include a book about some of the most important concepts in forecasting and statistical modeling, titled *Using Econometrics: A Practical Guide,* by A.J. Studenmund. This book delves to a greater degree into issues of forecasting, including multicollinearity, autocorrelation, heteroskedasticity, and countless other fundamental topics.

My second recommendation is of a more practical nature and is titled *A Gentle Introduction to Stata,* by Alan C. Acock. The reason I cite this work as a practical reference is because Stata is a statistical software package that can be used for analysis and forecasting. The book is as gentle as it is titled, and it makes the Stata software package, which is used by the Fed, public corporations, and forecasting firms like my own, easy to use.

I recommend these books because they are good starting points.

CHAPTER 8

WHEN TO CREATE DATA

You don't always face an analytical project with the data you want. Sometimes you're just stuck with what exists. But sometimes you have more power than that. You may actually be able to request that data be created.

There's a great line from the movie *The Untouchables* that makes me think about creating data. The film, which stars Sean Connery and Kevin Costner, includes a scene where the two are trying to cobble together a team of G-men to take down Al Capone.

At one point, Connery recommends that he and Costner seek out recruits from the policy academy. He offers up a kernel of wisdom to Costner, noting, "If you're afraid of getting a rotten apple, don't go to the barrel. Get it off the tree."

I think of this line all of the time when I'm working with data. Often, the exact and ideal data you are looking for may not exist. And that's when it's time to dream up what data you would like collected or created.

CHAPTER 9

HOW TO CREATE DATA

Data creation and collection can require Herculean effort and a lot of time. When talking about creating data, I often invoke an apropos Chinese proverb:

The best time to plant a tree was 20 years ago.
The second best time is right now.

This proverb hits the nail on the head, because to have a meaningful data series, it may take a long time to collect data that could prove most relevant. This is why the most important thing about creating data is to set the parameters and specifications right from the beginning. After all, you don't want to do it wrong because you don't want to collect data that may be unusable later — especially since you need to collect data for a long period of time to have truly meaningful implications.

And incorrectly setting the parameters of data collection, cohort, consistency, or other factors could render the data useless. More than once, I have had to scrap a data series after realizing we were not collecting the best data.

Data Creation Examples

One of the data projects that I am most proud of has been the creation of the MHI Business Activity Index (BAI) by Prestige Economics. This series is collected and published monthly. It is designed to reflect activity in the material handling industry, which is part of the U.S. supply chain.

We modeled some of the components on other activity indices. But we made a couple of modifications. We forced a ranking for participants, who had to respond that either they were seeing expansions or contractions in activity. In this way, the MHI BAI captures monthly changes in material handling business dynamics that does not allow for respondents to cop out and respond that month-on-month activity was flat.

Figure 9-1: MHI BAI by Prestige Economics

Material Handling Industry Business Activity

MHI	January 2019	December 2018	% Point Change	Growth Indication	Rate of Change	Trend in Months
MHI Business Activity Index	52	71	-19	Expansion	Deceleration	6
Capacity Utilization	53	59	-6	Expansion	Deceleration	13
New Orders	57	67	-10	Expansion	Deceleration	2
Shipments	61	48	+13	Expansion	From Contraction	1
Unfilled Orders	48	57	-9	Contraction	From Expansion	1
Inventories	44	33	+11	Contraction	Deceleration	2
Exports	45	56	-11	Contraction	From Expansion	1
Future New Orders	61	81	-20	Expansion	Deceleration	44

Prestige Economics, LLC   MHI Business Activity Index   Page 1

This is an important change from other activities, and it is important in order to ensure that the index is capturing changes at the margin. After all, if you ask most people how their business is compared to the previous month, you will likely get a high preponderance of respondents who say "no change."

We wanted to avoid that.

By forcing a choice between expanding and contracting activity, series readings above 50 indicate a majority of the respondents reported increased monthly activity, while readings below 50 indicate a majority of respondents reported decreased activity.

And we have been able to collect this data for business activity, capacity utilization, new orders, shipments, unfilled orders, inventories, exports, and future new orders (12 months in advance).

Data Changes

One of the most important attributes for data collection is to make sure that the cohort is committed. Similarly, you need to ensure that the data allows for opt-outs when relevant.

In the first few months of the MHI BAI, we did not have an opt-out for companies that did not have exports, inventories, or capacity utilization. Fortunately, with input from MHI, we adapted the survey very quickly and only lost a few months of data, in a survey that has been going for almost four years now. But if we had been slower to amend the survey, we would have needed to scrap a lot more data.

One of the first attempts at data creation has been the Prestige Economics Quarterly Benchmarking of risks, planned activity, and recession expectations from our corporate client cohort. The Prestige Economics Benchmarking survey is short, as is the MHI BAI. This is a critical piece to ensuring cohort participation. It is important to keep in mind that just because a survey is short does not mean that valuable implications cannot be teased out of the data. In fact, the MHI BAI is four full pages of graphs and text, and the Prestige Benchmarking fills a robust 20-page slide deck every quarter. The longer we collect this data, the longer the reports may be as the data may offer greater insights over time.

The MHI BAI and the Prestige Economics Benchmarking have been successful and effective. This is because they are anonymous and short — with committed participant cohorts.

Figure 9-2: Prestige Economics Quarterly Recession Risks

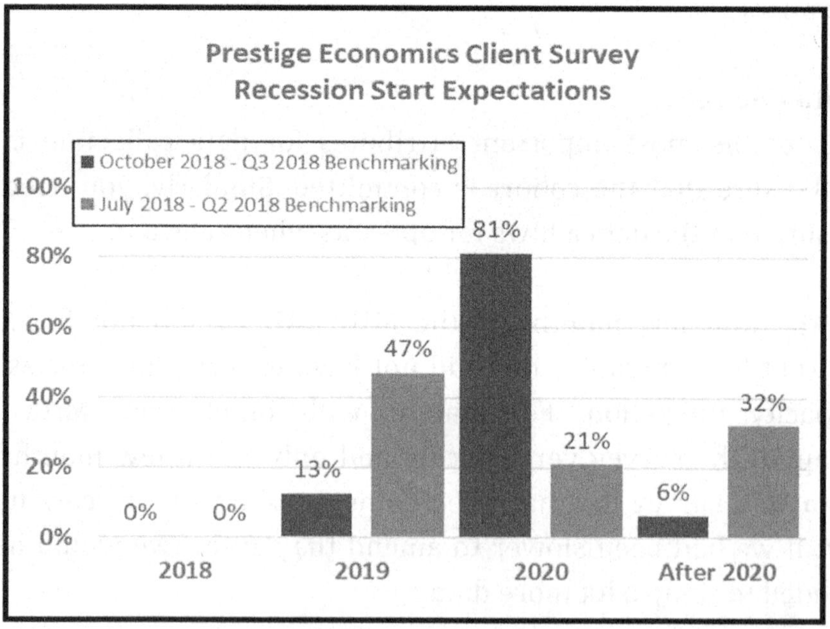

When Failure Strikes

One data collection project that failed was a Robot Activity Index (RAI) that we created with the *Robotics Business Review* (RBR). The Futurist Institute launched this index at the beginning of Q2 2018. The two-page report was based on a short quarterly survey of companies in robotics and automation. But the overall cohort was not very committed, and a low response rate fell during our three collection periods. After our third quarterly report, the index died, and we abandoned the project.

Sometimes this happens in data projects as well. You simply don't find enough value in the data collection and analysis to justify ongoing support. And especially with a data series that requires input from industry experts, we had to abandon the index as the number of participants was no longer statistically significant.

Figure 9-3: Robot Activity Index

RBR Robot Activity Index by The Futurist Institute

	Current Quarter	Previous Quarter	% Point Change	Growth Indication	Rate of Change	Trend in Quarters
New Orders For Current Quarter	83	100	-17	Expansion	Deceleration	3
Shipments For Current Quarter	83	100	-17	Expansion	Deceleration	3
R&D For Current Quarter	83	100	-17	Expansion	Deceleration	3
External Funding* For Current Quarter	67	33	+34	Expansion	From Contraction	1
Internal Funding* For Current Quarter	100	100	0	Expansion	No Change	3
New Orders 2018 versus 2017	83	100	-17	Expansion	Deceleration	3
Shipments 2018 versus 2017	83	100	-17	Expansion	Deceleration	3
Funding 2018 versus 2017	83	86	-3	Expansion	Deceleration	3

*Only if applicable.

The Futurist Institute RBR Robot Activity Index Page 1

Qualitative Data

The MHI BAI, RAI, and Prestige Quarterly Benchmarking have all been attempts to capture a mix of quantitative and qualitative data. Another attempt to collect data has been *The Robot and Automation Almanac.* But this is a collection of essays and perspectives. In that respect, it is more of a qualitative data project rather than a quantitative one.

The almanac fits with the other data creation projects we have spearheaded because it is an attempt to capture human intelligence-derived data, which is known in intelligence circles as HUMINT. Like the MHI BAI and Prestige Benchmarking, it has also been successful. We did our first almanac in 2018 and the second in 2019.

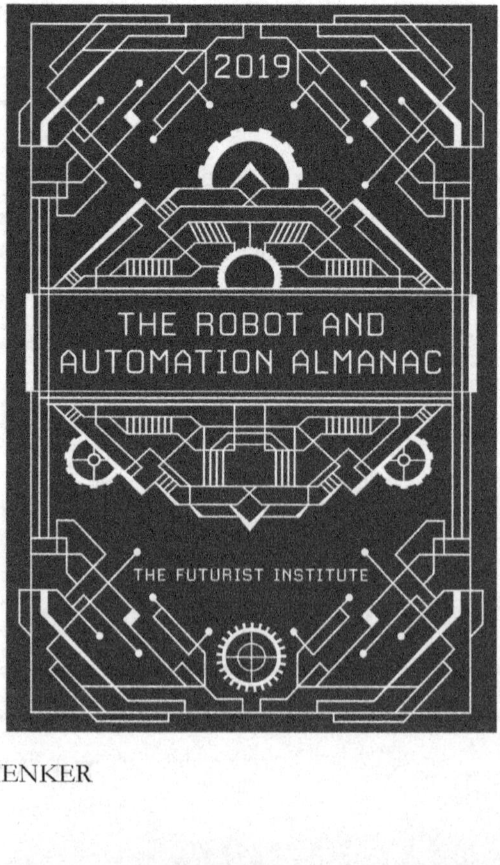

Summary of Good Creation Practices

Here is a summary of the important things you need to keep in mind when you create data, as I have noted in this chapter:

- **Try to focus on getting the data you want from the start to avoid having to reboot your data collection process.**
- **If you need to make a course correction in the data collection process, make sure to do so as quickly as possible. Any changes to your data series could invalidate previous data collected.**
- **Make sure that cohort participation is consistent.**
- **Keep the data collection for participants easy and short.**
- **Keep the data collected anonymous.**
- **If you aren't getting good cohort participation, you may need to scrap the data you have collected — and the entire project altogether.**
- **More data is better. Data has the most validity when it has been collected over a longer period of time.**

As I have noted in this chapter and the previous chapter, I would encourage anyone doing analysis to try and create the data that could best help them identify meaningful, actionable, and valuable insight. But it is important to remember that it is not an easy process — and that it may not work.

CHAPTER 10

CHECK THE FACTS AND SHOW UP

Quantitative data sets aren't the only data out there, as can be seen with *The Robot and Automation Almanac.* In fact, qualitative data can be even more important — especially if there are reliability issues with some of the quantitative data you are working with, like government data that is subject to major delays or revisions.

It's important to know that quantitative data should not just be taken prima facie as the entire story. It is important to perform a certain level of testing and sampling on data that you are using in your analysis.

This is essentially a form of audit or due diligence that you have to perform on your data. Sometimes this can be done by testing and sampling the data set you have.

But a big part of this includes showing up and checking the facts in person. If it's corporate data, you should find out more about the collection process — and see some of it yourself.

If you are dealing with macroeconomic or policy data, it is important to show up in person — and actually check the facts. This is part of my annual travel schedule. It's a mix of learning — one of the best practices I noted back in Chapter 6. But traveling around to meet with politicians and policymakers also serves a dual purpose.

It allows me to test my hypotheses with people who are in the know. Let me share a few stories to help illustrate the point.

The Incorrect Article

In the summer of 2012, a reputable magazine published an article that said a certain member of the European Monetary Union, which includes countries that use the euro as their currency, was planning to leave the Eurozone. Frustrated with the current dynamics in the group of countries that used the euro, this country was poised to leave. The story, if true, would have been bearish for the euro.

I was skeptical, but the stakes were high.

So, I booked a flight to Europe and set a meeting with the country's central bank. Fortunately, I have been meeting with central banks since 2004, and I had built a number of relationships that helped facilitate the meeting.

I sat down with one of the heads of that country's central bank, and I shared my views on the global economy and financial markets, as well as my expectations on the euro and the future of the European Monetary Union.

At the end of our talk, and as casually as I could, I mentioned that I was surprised to read the article about their country's thinking of leaving the Eurozone given my expectations that countries were unlikely to leave — and his similar sentiments about the prospects for the Eurozone. "Yes, I was too," he said.

He wasn't surprised that the story had leaked. He was surprised that even such a story had ever come into existence. It was ridiculous. But it had run in a major publication and foreign exchange markets were moving on the disinformation.

I walked out of that meeting with a bullish outlook on the euro, and I was proven correct over the following 18 months.

Knowing What to Write

In addition to meeting with foreign central banks, I also attend some events of the central bank — the U.S. Federal Reserve System, otherwise known as the Fed. Every year, I go to an annual event of the Atlanta Fed called the Financial Markets Conference. The first year I went, there were about 40 people, including Ben Bernanke, the then Chairman of the U.S. Federal Reserve System.

Now, about 300 people attend this event. It's an important event for senior global financial market leaders, policy makers, and a handful of market analysts. At the meeting in May 2018, the topic was the future of finance. Topics like machine learning, AI, and cryptocurrencies were on deck. And I overheard a number of conversations that conveyed confusion about Bitcoin, blockchain, Ethereum, and cryptocurrencies in general.

This inspired me to write my book *The Promise of Blockchain: Hope and Hype for an Emerging Disruptive Technology.* I figured if the paragons of the financial world were still a bit confused about the subject, I should write a book that elucidated some of the topics of greatest confusion. My clients appreciated the book. And SXSW asked me to do a formal book reading of it in March 2019.

OPECxpectations

For 15 years I have been going to OPEC meetings in Vienna, Austria. If you want to know what is going on in the oil market, you have to be there. There are dozens of reporters and analysts who attend. And there are even open sessions, where you can walk right up to the oil ministers of different countries and ask them questions directly.

Imagine that!

You can just go right up to their desks like they are the help desks of a library — and you can ask the most important oil ministers in the world anything! Of course, you have to be able to get into the meetings to begin with. And that's a trickier piece of business.

Perhaps the most important dynamic I have seen at an OPEC meeting in recent years was at the November 2016 OPEC meeting, where the decision was made for OPEC and non-OPEC members to coordinate on reducing their oil production in order to reduce global oil inventory overhangs and push oil inventories lower toward the five-year average level. This level was exceeded in large part due to a Chinese manufacturing recession that lasted from December 2014 until June 2016.

The logic behind the reduced OPEC production levels was that it would support prices, which it did — especially as Chinese manufacturing continued expanding until September 2018, when the inklings of a Chinese manufacturing recession came into focus.

The meeting was important because of the collaboration between OPEC and non-OPEC countries like Russia. But it was also important for another reason. Before OPEC and Russia and the rest of the countries involved got to the meat of a collaborative decision to reduce oil production, they unveiled the functionalities — including the UI/UX — of the new OPEC app in the opening session of the OPEC conference.

I snapped the picture below at that session, and it inspired me to speak more directly with my clients about the spreading need to adopt new and emerging technologies.

While the focus of the OPEC meeting was on the landmark OPEC and non-OPEC collaboration, my big takeaway was that OPEC was moving ahead of a number of major corporations with its technology push.

The Two Million Dollar Podcast

In stark contrast to the OPEC meeting takeaway that corporations needed to adopt technology was an interaction I had in 2017 in the startup world. I became an angel investor. Angel investors invest in, advise, mentor, and otherwise support startups.

One of the CEOs of a startup I met had a podcast. It was a four-minute-per-day podcast, and only the CEO had 7,000 email addresses on the mailing list.

Just to clarify, that doesn't mean all 7,000 opened the email every day. And it certainly doesn't mean that the people who opened the email clicked on the podcast. In fact, in my experience, a best-in-class open rate would be about 20 percent of all sent emails, which would be less than 1,400 emails. And about 20 percent of those would be a best-in-class click-through rate, or less than 300 people per day. And that's assuming the emails have absolutely best-in-class performance. Plus, the podcast was free. So there was no revenue generation.

The valuation the CEO was looking for was two million dollars. And it came through, with a 25 percent equity raise. In other words, tech fever was in full swing even in the private markets. And a four-minute-per-day podcast raised half a million dollars for 25 percent of the company.

I was shocked. I even wrote an article titled "Does Your Dog Need a Podcast?" I also shared this story with my clients and cautioned that it is symptomatic of some of the problems going on broadly with tech valuations of private and public companies in the past few years. And it is a trend that will likely end in tears.

Implications

Maybe you don't need to meet with policymakers to test your data hypotheses qualitatively. But chances are that some level of due diligence could add value to your analysis. This is what you need to figure out. How can you make sure that the data you are using represents what is actually going on? How can you check the facts? How can you be there — where the data is collected — and just behind a desk with data that may not be representative?

Data Analysis Best Practices

CHAPTER 11

NOT ALL DATA IS USEFUL

Not all quantitative or qualitative data is useful. This is why it is important to follow a proper data process that starts with asking the right questions and scoping the analysis you will need to do.

The truth is that there is more data now than could ever be useful. And the amount of data that is being created is increasing parabolically. This means that while a lot of useful data is being created, a lot of data that may not be useful is also being created.

There are three big reasons why data being created is not useful:

- **No one knows it exists.**
- **No one has found a use for it.**
- **It is noise that does not provide clarity.**

And these three problems exist for both quantitative data as well as for qualitative data. Both kinds of information can be found, for one reason or another, to be not very useful.

I became an economist in 2004 for Wachovia Bank, which was absorbed into Wells Fargo during the financial crisis. Within a year, I gave a talk at the Philadelphia Fed, which can be seen in the Figure 11-1.

At the time I became an economist, Fed press releases were short. An example thereof can be seen in Figure 11-2. The trappings of current Fed meetings did not yet exist. In 2004, there were no press conferences at quarterly Fed meetings, as became the practice under Fed Chairman Ben Bernanke.

And there were no press conferences at every single Fed meeting, as has been only recently announced by Fed Chairman Jay Powell.

Figure 11-1: Speaking at the Fed in an Era of Less Data[1]

In 2004, there was also no release of Fed member forecasts. These are the so-called dot plots, which show a visual distribution of the expectations for federal funds rate, inflation, and GDP expectations that are held by members of the U.S. Federal Reserve's Federal Open Market Committee. Today, those forecasts are released every quarter.

Figure 11-2: Fed Press Release in January 2004[2]

Federal Reserve Release

Press Release

Release Date: January 28, 2004

For immediate release

The Federal Open Market Committee decided today to keep its target for the federal funds rate at 1 percent.

The Committee continues to believe that an accommodative stance of monetary policy, coupled with robust underlying growth in productivity, is providing important ongoing support to economic activity. The evidence accumulated over the intermeeting period confirms that output is expanding briskly. Although new hiring remains subdued, other indicators suggest an improvement in the labor market. Increases in core consumer prices are muted and expected to remain low.

The Committee perceives that the upside and downside risks to the attainment of sustainable growth for the next few quarters are roughly equal. The probability of an unwelcome fall in inflation has diminished in recent months and now appears almost equal to that of a rise in inflation. With inflation quite low and resource use slack, the Committee believes that it can be patient in removing its policy accommodation.

Voting for the FOMC monetary policy action were: Alan Greenspan, Chairman; Timothy F. Geithner, Vice Chairman; Ben S. Bernanke; Susan S. Bies; Roger W. Ferguson, Jr.; Edward M. Gramlich; Thomas M. Hoenig; Donald L. Kohn; Cathy E. Minehan; Mark W. Olson; Sandra Pianalto; and William Poole.

2004 Monetary policy

Home | News and events
Accessibility
Last update: January 28, 2004

A Fed release calendar from 2004 was quite sparse, as you can see in Figure 11-3. Back then, even Fed meeting minutes were released with a six-week lag, such that the minutes of a meeting were not released until the next Fed decision had been published.

Today, that is not the case. Fed meeting minutes are now released three weeks after a Fed decision is made. That's half the time. And it means that minutes are released before the next Fed decision is made public.

You can see the more robust Fed calendar from 2018 in Figure 11-4. As I mentioned above, the new calendars will include press conferences at every single Fed meeting.

So why has the Fed increased its publications and access?

Figure 11-3: Sparse Fed Calendar — Q1 2004[3]

March 25, 2004
Minutes of Board discount rate meetings, December 15, 2003, to January 26, 2004

March 18, 2004
Minutes of Federal Open Market Committee, January 27-28, 2004

March 16, 2004
FOMC statement

February 5, 2004
Minutes of Board discount rate meetings, November 10 to December 8, 2003

January 29, 2004
Minutes of Federal Open Market Committee, December 9, 2003

January 28, 2004
FOMC statement

2004 Press Releases:

This has a lot to do with fallout from the financial crisis. The Fed has greatly increased its transparency so markets will have more time to respond to changes in policy, which would hopefully provide for less market volatility — and could hopefully prevent another financial crisis.

That may be a lofty goal, but increasing transparency and creating more data should hold significant value for financial market analysts. The question is, does it?

Figure 11-4: Packed Fed Calendar — Q4 2018[4]

2018 FOMC Meetings

January	30-31	**Statement:** PDF | HTML Implementation Note	Longer-Run Goals and Policy Strategy	**Minutes:** PDF | HTML (Released February 21, 2018)
March	20-21*	**Statement:** PDF | HTML Implementation Note	Press Conference **Projection Materials** PDF | HTML	**Minutes:** PDF | HTML (Released April 11, 2018)
May	1-2	**Statement:** PDF | HTML Implementation Note		**Minutes:** PDF | HTML (Released May 23, 2018)
June	12-13*	**Statement:** PDF | HTML Implementation Note	Press Conference **Projection Materials** PDF | HTML	**Minutes:** PDF | HTML (Released July 05, 2018)
Jul/Aug	31-1	**Statement:** PDF | HTML Implementation Note		**Minutes:** PDF | HTML (Released August 22, 2018)
September	25-26*	**Statement:** PDF | HTML Implementation Note	Press Conference **Projection Materials** PDF | HTML	**Minutes:** PDF | HTML (Released October 17, 2018)

There's a reason why this is the storyline in a chapter about how not all data is useful.

The Fed has gone to tremendous lengths to share its forecasts, its perspectives, its expectations, and its data. Even the length of Fed statements has increased significantly in the past 15 years, as you can see in Figure 11-5.

Figure 11-5: Fed Press Release in January 2019[5]

Press Release

January 30, 2019

Federal Reserve issues FOMC statement

For release at 2:00 p.m. EST

Share

Information received since the Federal Open Market Committee met in December indicates that the labor market has continued to strengthen and that economic activity has been rising at a solid rate. Job gains have been strong, on average, in recent months, and the unemployment rate has remained low. Household spending has continued to grow strongly, while growth of business fixed investment has moderated from its rapid pace earlier last year. On a 12-month basis, both overall inflation and inflation for items other than food and energy remain near 2 percent. Although market-based measures of inflation compensation have moved lower in recent months, survey-based measures of longer-term inflation expectations are little changed.

Consistent with its statutory mandate, the Committee seeks to foster maximum employment and price stability. In support of these goals, the Committee decided to maintain the target range for the federal funds rate at 2-1/4 to 2-1/2 percent. The Committee continues to view sustained expansion of economic activity, strong labor market conditions, and inflation near the Committee's symmetric 2 percent objective as the most likely outcomes. In light of global economic and financial developments and muted inflation pressures, the Committee will be patient as it determines what future adjustments to the target range for the federal funds rate may be appropriate to support these outcomes.

In determining the timing and size of future adjustments to the target range for the federal funds rate, the Committee will assess realized and expected economic conditions relative to its maximum employment objective and its symmetric 2 percent inflation objective. This assessment will take into account a wide range of information, including measures of labor market conditions, indicators of inflation pressures and inflation expectations, and readings on financial and international developments.

Voting for the FOMC monetary policy action were: Jerome H. Powell, Chairman; John C. Williams, Vice Chairman; Michelle W. Bowman; Lael Brainard; James Bullard; Richard H. Clarida; Charles L. Evans; Esther L. George; Randal K. Quarles; and Eric S. Rosengren.

Implementation Note issued January 30, 2019

Last Update: January 30, 2019

But for all its efforts to create and share more information, the Fed may have done so in a way that creates more data than markets can digest. After all, there are currently four forecasts of GDP from the Fed available at any single point in time.

The four different forecasts of GDP available right now from the Fed are as follows:

1. Annual year-on-year GDP expectations from the FOMC, which is part of a quarterly release.
2. Atlanta Fed's GDPNow, which is a frequently updated model of expected quarter-on-quarter annualized growth in the next release of GDP.
3. New York Fed's NowCast, which is a frequently updated model of expected quarter-on-quarter annualized growth in the next two releases of GDP.
4. St. Louis Fed's Economic News Index Real GDP NowCast, which shows expected quarter-on-quarter annualized real growth in the next releases of GDP.

In a poignant statement about Fed policy, Janet Yellen noted at Jackson Hole, Wyoming, in August 2016 that "Our ability to predict how the federal funds rate will evolve over time is quite limited."[6] So, does all the data help? Maybe not.

For all of its data and transparency, the Fed only has so much control over imbalances, policies, and behaviors that could cause another recession. This doesn't mean that all the data is useless. But it cannot do the one thing that market mavens wish it could do the most: offer predictive insights into the future of the federal funds rate and other interest rates.

It's important to consider that the Fed struggles with making good forecasts, even though it has almost limitless resources and a massive staff of well-educated and highly skilled economists and econometricians.

The takeaway for working with data is that sometimes all the data in the world may not be enough to help you answer the questions you need answered. In other words, as advertised, not all data is useful.

CHAPTER 12

IGNORING SOME DATA IS IMPORTANT

Since not all data is valuable, it is important to ignore some data.

When volumes of data seem limitless, so are the analytical possibilities. Without limits on data, there is a risk of analysis paralysis — when you are overwhelmed by the data and making decisions is difficult.

There are many times when this situation comes up and you are working with data. It happens in analyses of financial market and economic data. And it also happens in large panel data sets — and in massive corporate data set collection.

I have personally seen this on many occasions.

It is important not to get caught up in considering *all* data — just the data that actually matter. I confronted this kind of challenge early in my career, when I was analyzing oil markets. A client asked about the difference between two different weekly oil inventory reports, and which one should be trusted most.

One of the weekly oil inventory reports is produced by the U.S. Department of Energy (DOE). The other weekly oil inventory report is produced by the American Petroleum Institute (API). Even though these two reports purport to show the same data, they often show massively divergent dynamics.

And the difference in reliability can be judged on the oversight of that data collection. In my analysis of the two reports, I discovered that the DOE inventory report was subject to the rule of law — and that there could be criminal and civil penalties if the data provided are not accurate. The instructions for the DOE form include the following caution:

> The timely submission of Form EIA-803 by those required to report is mandatory under Section 13(b) of the Federal Energy Administration Act of 1974 (FEAA) (Public Law 93-275), as amended. Failure to respond may result in a civil penalty of not more than $2,750 per day for each violation, or a fine of not more than $5,000 per day for each criminal violation. The government may bring a civil action to prohibit reporting violations which may result in a temporary restraining order or a preliminary or permanent injunction without bond. In such civil action, the court may also issue mandatory injunctions commanding any person to comply with these reporting requirements.[1]

Even the actual weekly crude oil stocks report form EIA-803 is headed with a stern warning that underscores the importance of providing data — true and complete data:

> This report is mandatory under 15 U.S.C §772(b). Failure to comply may result in criminal fines, civil penalties, and other sanctions as provided by law. For further information concerning sanctions and data protections, see the provision on sanctions and the provision concerning the confidentiality of information in the instructions. 18 U.S.C. §1001 makes it a criminal offense for any person to knowingly and willingly make to any Agency or Department of the United States any false, fictitious, or fraudulent statements as to any matter within its jurisdiction.[2]

The risk of committing a criminal offence, or being subject to civil or criminal fines and penalties, should be enough to ensure that the weekly DOE report forms are completely and accurately completed.

The API is an important industry group, but its oil inventory reporting forms do not have the backing and power of law. For this reason, I have encouraged my clients to ignore the API data. After all, there is no requirement to comply and provide weekly stocks data, nor are there any repercussions for providing false information in weekly API data collection.

This may be an extreme example of why it is important to ignore some data. But whenever you are looking at different data, it is important to consider if it is useful.

You don't have to squeeze every potential variable into financial models you are building for market or corporate data. But you do need the ones that make the model more accurate.

This is also something I have seen in my analysis of the Chinese economy.

There are many Chinese economic data reports, but not all of them are valuable — or worthy of consideration. Of course, the same can be said of the economic reports of many countries. But some view Chinese government data with extreme skepticism.

Regardless of whether you trust or distrust official Chinese government data, one Chinese economic data point that is beyond reproach is the Chinese Caixin Manufacturing PMI. After all, this is a report that is not collected by the Chinese government. It is not an official data release. Plus, it is a survey of small and medium-size manufacturers, which are less directly influenced by Chinese government policy than large manufacturers. Data that is above reproach is hard to come by. And that makes it the most worthy of consideration — and that makes it possible to ignore other Chinese economic data.

No matter how much data is collected or produced, do not be afraid to ignore and disregard data that do not improve the validity or predictive nature of your analysis. By sticking to value-add data, it will make it easier to get buy-in for your analysis.

CHAPTER 13

GETTING BUY-IN IS CRITICAL

The most important part of any analysis is that it is actionable — and that any examination and consideration of the data and analysis is sufficient to gain buy-in. Whoever your stakeholders are, they need to believe in your capabilities, they need to comprehend your logic, and they need to understand the implications of your conclusions.

You can have the best analysis in the world, but you need people to support your assertions. This requires sound analysis, but it also requires effective communication.

In a world of exponentially increasing data, it will be increasingly important to cut through the noise of limitless available data. It will be equally important for analysts to help guide their managers, stakeholders, and colleagues through a thickening fog of data.

How to best do this is the subject of the next section of this book.

Best Practices for Communicating Data

CHAPTER 14

DEPICTING DATA: SHOW, DON'T TELL

The most important way to get buy-in when you discuss data is to show people things rather that explain ad nauseum to the point where they lose focus and attention.

It's important to remember that:

- **People love graphs.**
- **People love images.**
- **People love pictures.**

Long before Instagram, there were people's slideshows of their trips and vacations. Even in statistics and economic professions, people would rather see charts and tables and images rather than just read words.

The more visual the image, the more it is likely to reach an audience, whether you are giving a speech, leading a meeting, or talking to a manager. After all, charts with lots of data — or rather too much data — are even called *eye charts*, because they are tough to read and unpleasant to look at.

Explaining Multivariate Regression Analysis

On more than one occasion, I have had to explain multivariate regression analysis. It's a more difficult concept for people not well versed in statistics to grasp easily. But as someone who creates, analyzes, and communicates about data, it is my job to make things easy to understand.

I like to explain the difference between univariate analysis and multivariate analysis using a series of three images. In Figure 14-1, you can see the first of these three images. I usually explain that this image represents a univariate analysis. The input is on the right, the output on the left. There is one input to the fried egg: an egg.

Figure 14-1: Univariate Analysis with One Variable[1]

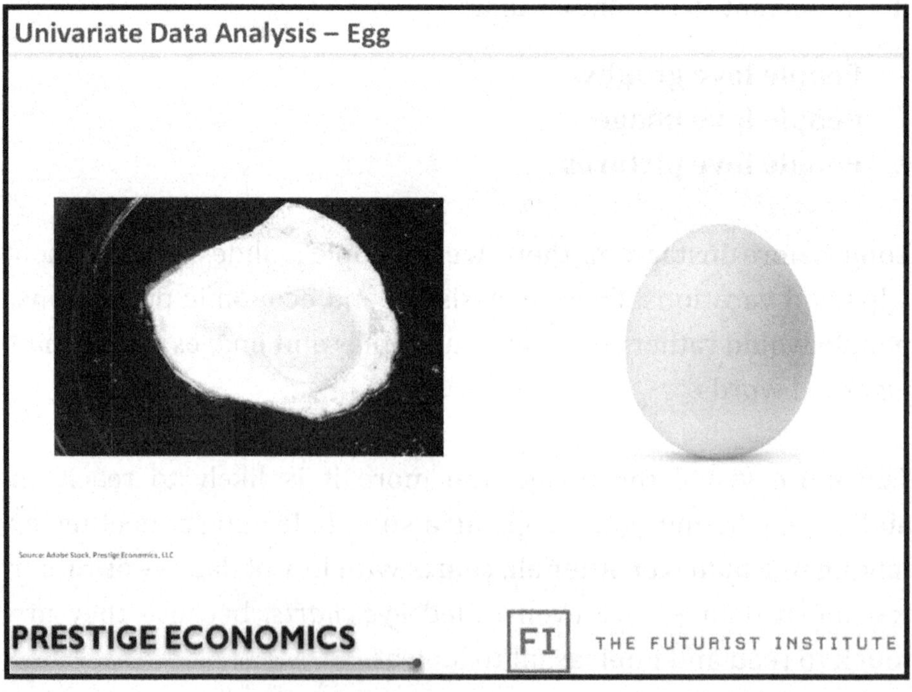

This is why you can easily see how similar the input and output are.

I then like to show Figure 14-2, which shows a loaf of bread as the output and the egg as the input. This is what a multivariate analysis is like. After all, you might not easily recognize the input (in this case, an egg) in the output data (which here is bread). Nevertheless, the input is part of determining and predicting what the output will be. The input is one of the ingredients of an analysis thereof. But it is not the one and only ingredient.

For the third and final image of this series to explain multivariate analysis, I then share Figure 14-3.

Figure 14-2: Multivariate Analysis is Different[2]

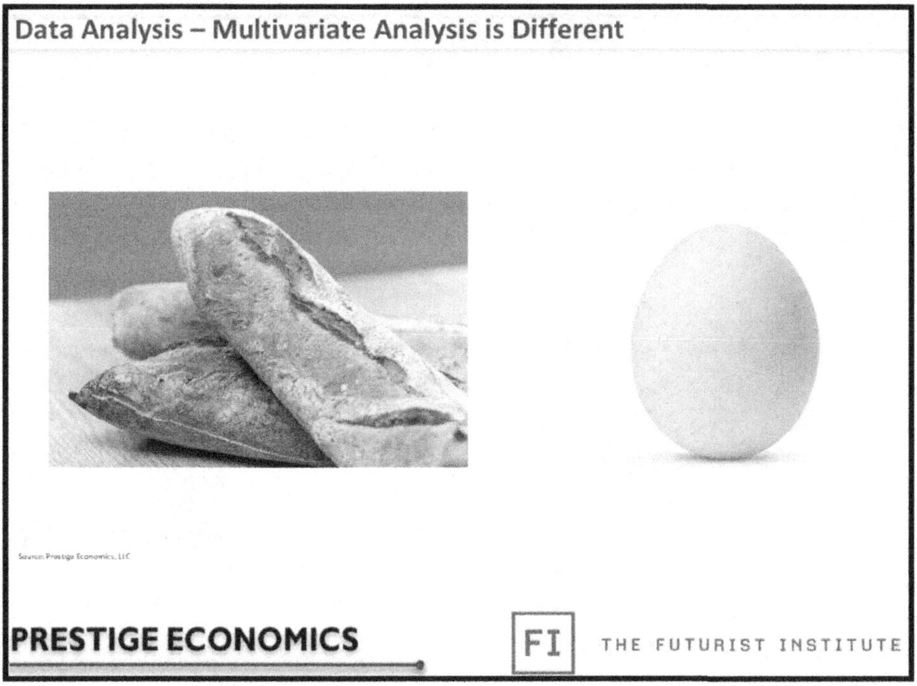

This image shows four different inputs required to create the output. I explain that this is what a multivariate analysis is like. It is a series of statistical inputs required to create a comprehensive analysis. And even if each of the ingredients does not completely explain what the output looks like, it is the mix of those ingredients working together that creates the output.

In the same way that bread is not just grain, or eggs, or salt, or milk, a multivariate regression analysis is a way to statistically predict a specific outcome by using a number of different inputs that together can logically and statistically explain the output.

This kind of visual imagery has been very effective.

Figure 14-3: Multivariate Analysis with Four Variables[3]

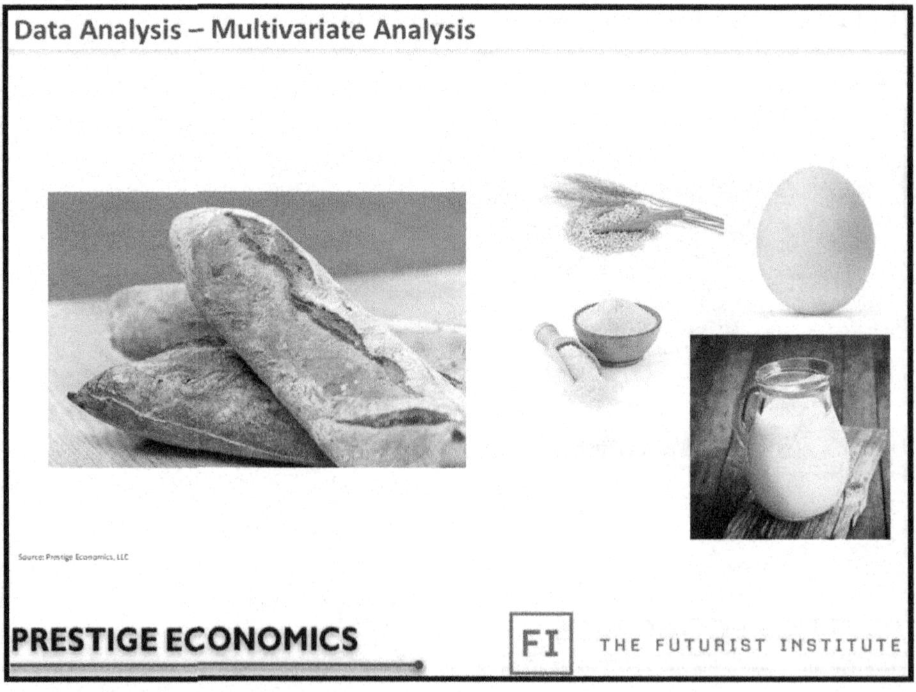

Stressing the Importance of More Data

In addition to frequently discussing the difference between univariate and multivariate analysis, it is also common for me to discuss the importance of creating and collecting more data. As I have noted elsewhere in this book, more historical data provides a more comprehensive picture of the statistical relationships and situation being analyzed.

There are two images I like to use when making this point. The first, in Figure 14-4, shows a set of data. If this were data over time, which is otherwise known as time series data, a statistical look at this series might lead you to believe that this data is moving sideways — and had been falling.

Figure 14-4: Limited Historical Data

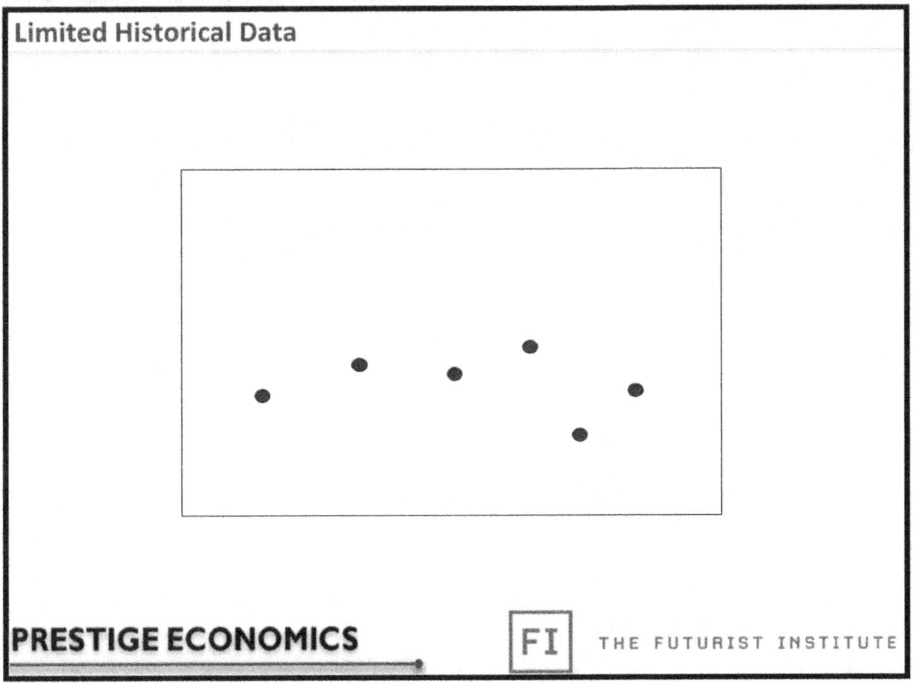

In order to challenge this view, and stress the importance of collecting more data, I then show the data points at the end of a much longer time series of data, as shown in Figure 14-5. Although the recent data points have been moving relatively sideways, the other historical data had been moving up and to the right in a very strong growth trend.

This is the graphical depiction of how the trendline in Figure 14-5 may be more important than the headline one might derive at first blush by just looking at Figure 14-4. In general, trendlines trump headlines. And long-term trends — and more data, in general — are more helpful in deriving useful implications from data analysis and for making predictive forecasts.

Figure 14-5: More Extensive Historical Data

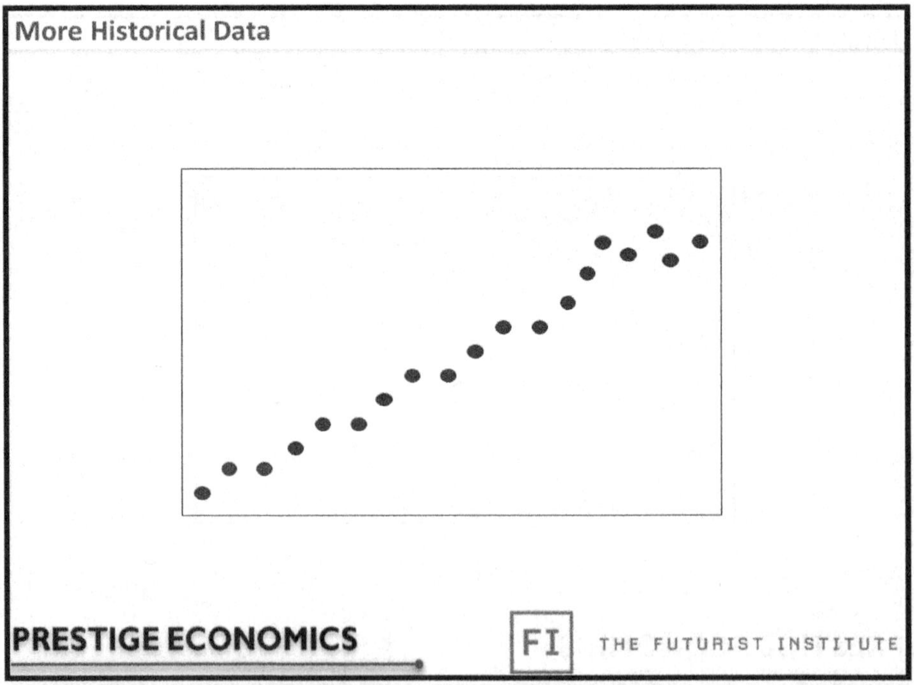

I always struggle with a need for more data in forecasting projects. You need more data, because the more data you have, the more accurate your forecasts are likely to be. This is because more data will usually include more information that can be useful in making predictions in a more robust series of scenarios.

But the data you are looking at also has to be good.

Bad Data

One of the groups I do forecasting work for is MHI; it's the trade association and industry group for the material handling industries, which represent the companies that provide goods and services to companies in the U.S. supply chain.

One of the critical data series that I forecast for the group is the U.S. material handling equipment manufacturers (MHEM) data series, which is provided as part of official government data on durable goods orders. Even though this data is historical, it has very often been subject to very large historical revisions of +/- 20 percent in any given year.

That's a lot!

In my investigation of the problems with the data and revisions, I discovered that the data we had been working with had other big problems. And I decided to share the outcome of our deeper analysis with the members of MHI. To do so, I chose some imagery that would resonate clearly so that the absurdity of the data would be clear — and so that the members would fully understand the situation.

There were two big problems with the official government MHEM data set, which included four NAICS codes associated with the inputs:[4]

NAICS 333921 Elevators and Moving Stairways
NAICS 333922 Conveyors and Conveying Equipment
NAICS 333923 Overhead Traveling Cranes, Hoists, and Monorail Systems
NAICS 333924 Industrial Trucks, Tractors, Mobile Straddle Carriers, and Stacker Machinery

The issue was that some critical industry data was missing from the official government series. But there were also other data that did not quite fit. The first image I showed was Figure 14-6.

Figure 14-6: Imperfect Data — Missing Elements[5]

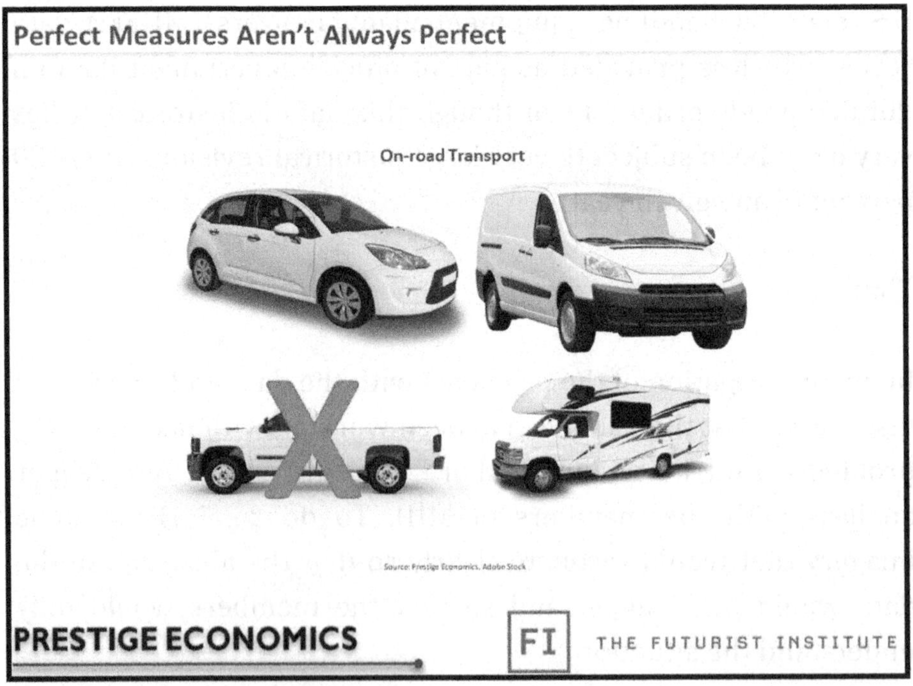

This shows an image of on-road vehicles but excludes pickup trucks. I did this to make the comparison that some of the most important parts of material handling — like racks used in warehouses and distribution centers as well as automation solutions — were not part of the official MHEM government data.

As a follow-up, I showed Figure 14-7, which includes on-road vehicles and excludes pickup trucks, yet somehow also includes trains. This was to highlight the odd inclusion of elevators and moving stairways as part of the official MHEM data, which are not really what we would consider material handling data. They are about as much as material handling data as a train is on-road transport. I mean a train is on the ground, but it doesn't fit.

Figure 14-7: Imperfect Data — Wrong Elements Added[6]

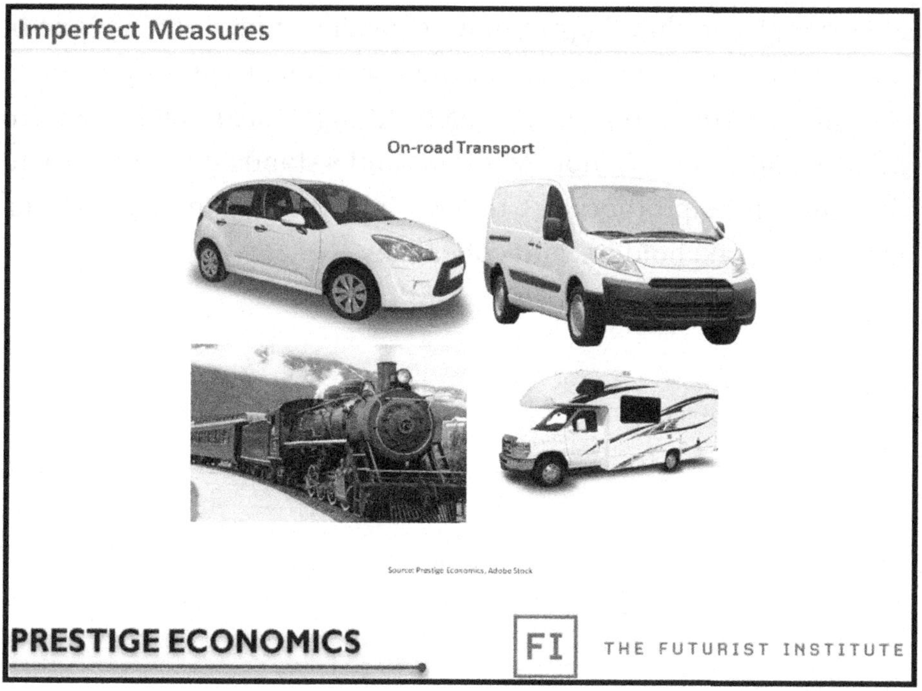

This presentation proved effective to making the audience quickly and clearly grasp some of the data challenges at hand. Unfortunately, it might quite literally take a multinational trade agreement or act of Congress to change these classifications that no longer match current industry conditions. So, we are stuck with the imperfections of the data. But at least everyone knows the situation.

Sometimes when you explain a data relationship or situation with images, the message may be unpalatable. But your job isn't to fix the problem — especially if it isn't fixable. But it is your job as someone who works with data to make everyone aware of the challenges, risks, opportunities, and value of the data you are working with.

My examples in this chapter may seem silly at first. But anyone in the world would quickly grasp the concepts at hand. And that's the point. Your job is to make things clear and easy to understand. After all, not everyone understands data. But if you do, you can help others get there. It's your duty to your clients, managers, stakeholders, and colleagues.

CHAPTER 15

TELL STORIES

If you want buy-in, you have to tell stories. Stories that resonate with people. Stories that make sense.

Just as data analysis needs to follow a process, and analysis needs to be logical, so does the communication of results, expectations, and forecasts. As a further extension of the recommendation to visually depict data relationships and issues, I would further recommend using a mix of images and stories to convey meaning.

One expert I know in the intelligence community is fond of saying that the plural of anecdote is data. Making sure you tell anecdotes that make sense and resonate can provide validity in the absence of data. This is exactly what I do in my futurist presentations. It's also a core aspect when I train consultants, executives, analysts, and strategists to become futurists.

It is important to use historical context in combination with the data of the present to build a vision of the future. In this chapter, I provide a few examples of how I use stories and imagery to make predictions that are designed to resonate.

Blockchain

One of the hottest new technology topics in recent years is blockchain. A story I tell draws on ancient times and is related to the ancient burning of the Library of Alexandria. In fact, it is the cornerstone of my book *The Promise of Blockchain*.

A central point of failure is a single point of failure that threatens the existence of an entire system because of centralization. A centralized exposure and concentration in the system threatens the entire system. A great example of a central point of failure was the burning of the ancient Library of Alexandria by Caesar during the Alexandrian Wars.

Figure 15-1: The Fire at the Library of Alexandria[1]

Up to half a million scrolls of papyrus perished, with massive amounts of knowledge from the ancient world, when the world's leading depository of records and writings burned.

Against a backdrop of this ancient story of loss is the new and emerging modern blockchain technology, which offers the promise of record-keeping redundancy in a decentralized framework. It is something new that can solve a problem as ancient as time itself. And wrapped up in this ancient imagery is a vision of a future in which the kind of institutional loss of knowledge represented by the fire at the Library of Alexandria can be prevented.

The Future of Online Education

Another futurist story I tell is about changing dynamics related to online education growth. I could share an image of a large classroom studio. But the most striking image is of a green room.

Figure 15-2: The Online Education Green Room at UT Austin[2]

A green room is where on-camera talent goes to prepare to be on camera. The image in Figure 15-2 is of a green room at UT Austin. It was once a classroom, but now it is where professors relax and put on makeup before they record online courses. The fact that the production organization has made room for a green room indicates the imperative of online education — and the willingness to sacrifice a classroom.

The Past of Work

When I talk about robots, one of my favorite things is to share the image in Figure 15-3. Work has always been changing. Proof of this is the fact that the family name Smith is the most common in the English-speaking world, yet there are few smiths remaining in the world today.

Figure 15-3: The Past of Work[3]

The story here is that the dynamics of a changing labor market in the 1800s were tectonic. The dynamics of a world with limited communication, limited access to education, and limited retraining potential stands in stark contrast to a world with ever-expanding online education, retraining, and remote working dynamics. The takeaway is that while the future of work will involve significant and rapid change, we are vastly better prepared and equipped today to manage those changes than a blacksmith was in the 1800s.

The Future of Work

As a final story I tell about the future, I underscore that everyone knows about the coming changes with robots, automation, and artificial intelligence. When I want different groups of executives, professionals, or students to tangibly understand the technological frenzy that currently pervades our *Zeitgeist*, I share the image in Figure 15-4.

Figure 15-4: The Future of Work[4]

This image was hanging in the airport of Amarillo, Texas, in mid-2017 — almost two years ago. The story here is that awareness about the coming future of work dynamics pervades even smaller cities with more agrarian economic attributes.

Telling stories is a highly effective way to share a vision. It is a way to take limited data points and extrapolate them into the future in a way to increase believability and buy-in. For this very reason, it is important to be careful what stories you tell.

If the stories you tell with data, anecdotes, or historical allegory are too extreme — or if you discuss technology that is nowhere near adoption or even minimum viable product — you may lose credibility. And people will think you are weird.

It's OK to be a quirky data scientist, economist, analyst, or forecaster. In fact, there is some expectation thereof. But it's never OK to come off as weird. It will undercut your message and make you an ineffective communicator.

CHAPTER 16

DON'T BE WEIRD

Talking about the future can be a double-edged sword. On the one hand, it can be exciting and interesting. On the other hand, there is an opportunity to lose credibility. And in futurist circles, there is an unfortunate tendency toward exaggerated eccentricity.

I've seen speakers with futuristic looking sci-fi outfits and glasses talk about going to colonize Mars, reprinting our entire bodies, and building cities under the sea as if these things will happen tomorrow. While these kinds of topics make for great sci-fi movie plots, their reality would likely be far in the future — if ever. And focusing too much on these kinds of topics can reduce the credibility of a data scientist or analyst speaking to professionals or executives.

I built The Futurist Institute to train others to become futurists — to incorporate new and emerging technology risks and opportunities into their strategic planning. I did not create it to foster a community of future-obsessed fanboys. But there are some futurists who fall into that category. And they are not taken seriously as professionals.

The media works itself up into a hype frenzy over every single technological advancement that could do something miraculous. The cost doesn't matter. The practicality doesn't matter. All that matters is fascination with the topic, clickbait potential, and hype.

The most important thing I could recommend to any forecaster, predictive data analyst, economist, or futurist is to help your stakeholders, your managers, and your colleagues see your point of view — and to help them find it reasonable. You may be wrong, and in fact you may well be wrong in *many* cases.

But being an analyst isn't about always being right. It is about following the most responsible process, performing due diligence on the quantitative and qualitative data available, and then sharing your insights in a reasonable way designed to resonate with stakeholders. You may be wrong, but at least you will be respected.

Of course, if you show up at a meeting carrying one of Elon Musk's blowtorches from The Boring Company while wearing futuristic sci-fi clothes, and you begin musing about your desire to telecommute off-planet for when you relocate to Mars, you may again be wrong — but you are much more likely to lose the respect of your colleagues than if you stick to technologies that could have a commercial impact in the more medium term.

The big takeaway here is to try and stick to discussing technologies that have achieved minimum viable product and are on the verge of commercialization — or which will be close to commercialization soon.

CHAPTER 17

THIS TIME IS NEVER DIFFERENT

An unfortunate but all-too-common aphorism among market mavens, traders, and economists is that *this time is different*. But I am pleased to say that I know of one bank where analysts get fired if they utter those words. Their mantra is *this time is never different*.

It is a sentiment that I also subscribe to.

Whenever there are financial market, economic, or data dynamics that look truly unique, it is often the case that there are historical allegories that are worthy of consideration. For someone looking at data, it is important to find patterns or dynamics that mirror past historical events.

This is true whether you are looking at sales data, logistics and transport data, economic indicators, or financial markets. The aphorism *history never repeats itself, but it often rhymes* seems most apropos on this subject. In fact, this is the notion behind statistical analyses of historical data. It is the search for predictive patterns.

This is also the reason why more robust data series are of the greatest value in performing analyses. They offer the greatest potential to capture historical events in order to better predict the future. I have discussed this concept in previous chapters.

When discussing cryptocurrencies in my book *The Promise of Blockchain,* I use a mix of imagery, a story, and the notion that this time is never different to elucidate a technology dynamic that has been confusing. In my opinion, bearer bonds, like the one in Figure 17-1, are the paper equivalent of digital cryptocurrencies. They can be used legitimately but are often a preferred medium of exchange of bad actors.

Figure 17-1: Bearer Bonds[1]

1,000,000

1,000,000

7⅝%
TREASURY BOND
OF 2002-2007
Dated February 15, 1977
Due February 15, 2007
CUSIP 912810 BX 5
Redeemable on call on and after
February 15, 2002
CIRCULAR No. 4-77

1586F

THE UNITED STATES OF AMERICA

FOR VALUE RECEIVED PROMISES TO PAY TO THE BEARER THE SUM OF

ONE MILLION DOLLARS

ON THE DUE DATE, AND TO PAY INTEREST ON THE PRINCIPAL SUM FROM THE DATE HEREOF, AT THE RATE SPECIFIED HEREON. THIS BOND AND INTEREST COUPONS ARE PAYABLE AT THE DEPARTMENT OF THE TREASURY, WASHINGTON, D.C., OR AT ANY FEDERAL RESERVE BANK OR BRANCH. THIS BOND IS ONE OF A SERIES OF BONDS, AUTHORIZED BY THE SECOND LIBERTY BOND ACT, AS AMENDED, ISSUED PURSUANT TO THE DEPARTMENT OF THE TREASURY CIRCULAR REFERRED TO HEREON. ALL OR ANY OF THE BONDS OF THIS SERIES MAY BE REDEEMED, AT THE OPTION OF THE UNITED STATES, ON AND AFTER FEBRUARY 15, 2002, AT PAR AND ACCRUED INTEREST, ON ANY INTEREST DAY OR DAYS, [illegible] MONTHS' NOTICE OF REDEMPTION GIVEN IN SUCH MANNER AS THE SECRETARY OF THE [illegible] PRESCRIBE. IN CASE OF PARTIAL REDEMPTION THE BONDS TO BE REDEEMED WILL BE [illegible] SUCH METHOD AS MAY BE PRESCRIBED BY THE SECRETARY OF THE TREASURY. FROM THE [illegible] DESIGNATED IN ANY SUCH NOTICE, INTEREST ON THE BONDS CALLED FOR [illegible] CEASE. THE INCOME DERIVED FROM THIS BOND IS SUBJECT TO ALL TAXES IMPOSED UND[illegible] INTERNAL REVENUE CODE OF 1954. THIS BOND IS SUBJECT TO ESTATE, INHERITANCE, GIFT OR OTHE[illegible] TAXES, WHETHER FEDERAL OR STATE, BUT IS EXEMPT FROM ALL TAXATION NOW OR HEREAFTER IMPOSED ON THE PRINCIPAL OR INTEREST HEREOF BY ANY STATE, OR ANY OF THE POSSESSIONS OF THE UNITED STATES, OR BY ANY LOCAL TAXING AUTHORITY. THIS BOND IS ACCEPTABLE TO SECURE DEPOSITS OF PUBLIC MONEYS. IT IS NOT ACCEPTABLE IN PAYMENT OF TAXES.

WASHINGTON, D. C., FEBRUARY 15, 1977.

SECRETARY OF THE TREASURY

1,000,000

ONE MILLION DOLLARS

1,000,000

Never Different

The notion that *this time is never different* applies just as equally to corporate data as it does to the nature of technological trends as well as economic and financial market data. Predictive data analysis of all kinds hinges firmly on the notion that this time is never different.

After all, if this time truly is different, then analyzing historical data would be worthless. There is no predictive value if the past offers no insight into the future. For those looking at data, it is critical to keep this in mind. Because no matter how advanced technology becomes, if a robust set of historical data is not representative of all potential dynamics and realities, then all the analysis in the world would prove irrelevant and worthless.

We know that historical data will hold increased future potential to predict client, financial market, and technological developments. As such, this time can never truly be different.

CHAPTER 18

BEING WRONG

As someone who works with data professionally, it is not uncommon for me to be wrong. It's inexcusable to be wrong about historical data — unless, of course, you are the government and produce data that is subject to revisions. But it is a normal part of being a data analyst to see that the future does not align perfectly with your forecasts.

Almost every forecast you make is likely to be wrong to a certain degree. This is a tough reality for a forecaster or analyst to swallow. But the goal is to get better at forecasting over time. This harkens back to the notion of testing and retesting data and forecasting results as part of a fundamentally sound data process that I prescribed in Chapter 4.

With enough data and enough testing, the hope is that the accuracy of any predictive analysis would be more effective. This is the goal of any analyst — to work tirelessly to improve forecast accuracy. But it's also part of the job to know that you will be wrong — often.

Dangerous Data

CHAPTER 19

MISUSE OF DATA

Misuse of data is a real problem. And how companies use their own data is often a bigger problem than how outsiders use such data.

Along those lines, there is a story about a businessman who summons three people to his office to review his financial accounts. The first is an accountant. The businessman asks the accountant what his business numbers mean. The accountant discusses profit margins, earnings, and different financial ratios. The businessman then summons a mathematician and asks the same thing of the mathematician, who talks plainly about the trends in the figures. Finally, the businessman summons a statistician to his office. The businessman again poses the same question. This time, however, the statistician pauses for a minute. Then he closes the door before he turns to the businessman and says, "What do you want the numbers to mean?"

There is some truth to this story, as there is to U.K. Prime Minister Benjamin Disraeli's statement about three kinds of lies: lies, damned lies, and statistics.

The problem with statistics and the reason they fall into Disraeli's list of lies is that they can be misused in the wrong hands.

Too Dumb

One of the most egregious situations I encountered involving a misuse of data was one in which an analyst at another firm was intentionally misleading his clients.

At a public conference, I saw him share one particular graph with a room full of executives, analysts, and investors. He was trying to show a correlation between the prices of two commodities, with one leading and the other lagging.

The problem with his depiction was that he had manipulated the axes of a graph in order to make his case. Having previously done a similar analysis myself, I knew that normalized axes would actually show that the leading commodity was actually the lagging one.

Believing this to be an honest mistake, I approached him about the data and graph after his talk. I politely pointed out that the axes may be showing something unintentional. And I noted that the axes should be adjusted to more properly reflect reality — which was the opposite of what he was trying to show.

He responded, "Our clients are too dumb to know the difference."

To treat one's clients that way — to manipulate them as well as think and speak so lowly of them — is shocking. That, my friends, is a *schande*. It is a clear-cut case of fraud.

Of course, this isn't the only time in my career when I have seen people manipulate data. While doing so in a public venue in front of clients you are defrauding is exceptionally egregious, having a healthy skepticism about data is critical. There have also been other occasions when I have seen data sets used in certain ways to exclude or include certain subsets in order to make very specific arguments.

Moral Hazard

One major risk in the world of forecasting is something called moral hazard. This means that your personal biases, opinions, or interests may inadvertently impact your opinions.

In essence, it is a conflict of interest that impacts the analyst.

And it's something important to watch for.

In a recent presentation, I asked a room full of government analysts the following question: *What the #1 job of a forecaster in the private sector?*

As I expected, they answered that it was to produce accurate forecasts. I pointed out to them that in my experience, the #1 job of any private sector analyst is to *not get fired.*

We often forget that people sometimes have bigger agendas that dominate their job performance. And keeping a job is likely to be far more important to an individual than pushing back hard against forecasts you disagree with — or updating your forecasts rapidly.

Imagine if you worked for a large financial institution and a client bought a yard of Japanese yen (JPY) based on your forecast that JPY would rise to 100 against the dollar. A yard, by the way, is a billion of something in the financial world. So, someone just bought a billion JPY on your recommendation.

Now, imagine that the very next day something happens and you need to change your forecast to show JPY weakening to 120 against the dollar. That's a 20 percent different on your forecast. And it means a potential loss of 200 million JPY for your client.

Now, how quickly would you want to tell your client that they are going to lose 200 million yen based on your recommendation from yesterday? More importantly, how quickly does the salesperson who accompanied you want your forecasts updated? You might like to think they would want them to be updated right away. But with that much money on the line, there might be massive pressure for the analyst to save face — and not get fired.

Takeaways

The most important thing to know as a data analyst is that trust is a fragile thing. Like someone who exaggerates or lies on their résumé. Some of the ways in which people most commonly manipulate data is with axes and graphs that appear to show relationships that are, in fact, nonexistent. This is the kind of data misuse that leverages the power of imagery for malicious purposes. Similarly, although somewhat less nefarious (maybe), are conflicts of interest. These too can impede the independence of a forecaster, causing a breakdown of trust — and having significant negative professional and business impacts.

CHAPTER 20

THE CURRENT STATE OF FAKE NEWS

People mean different things when they say the words *fake news.* But whatever people mean when they reference fake news, it is important to remember that this time is never different.

False advertising, propaganda, and spreading disinformation are not new. Snake oil salesmen have been around as long as snakes.

But the medium of modern fake news is the internet. That part is new. Nevertheless, the origins of the rise in the tide of modern fake news are actually more innocuous than you might think. After all, the corporate world has been using the internet for lead generation and sales from the very beginning. The new dynamic is that social media websites like Facebook, Twitter, Pinterest, Snapchat, and Instagram have created massive opportunities for companies to subtly (although sometimes not so subtly) generate leads and sales.

But as the saying goes, if man can make it, he can break it. And fake news is just one way that bad actors have been wreaking havoc in the world online.

Companies have been using social media and the internet to fill their marketing and sales pipelines using content marketing. This is a strategy that involves using substantive content to gain attention and generate business leads. Content, in this sense, often includes podcasts, articles, vlogs, and even books that are designed to build a following and create brand affinity.

The logic of generating content on the internet is that it will draw clients to your firm. You get what sales people call inbound leads. There is a website called Hubspot that has been a leading pioneer in the world of content marketing. And the term for these content-generated inbound leads is called inbound marketing. Hubspot even has a catchphrase about this on their webpage, noting that inbound marketing is "The best way to turn strangers into customers and promoters of your business."[1]

Figure 20-1: Client Funnel[2]

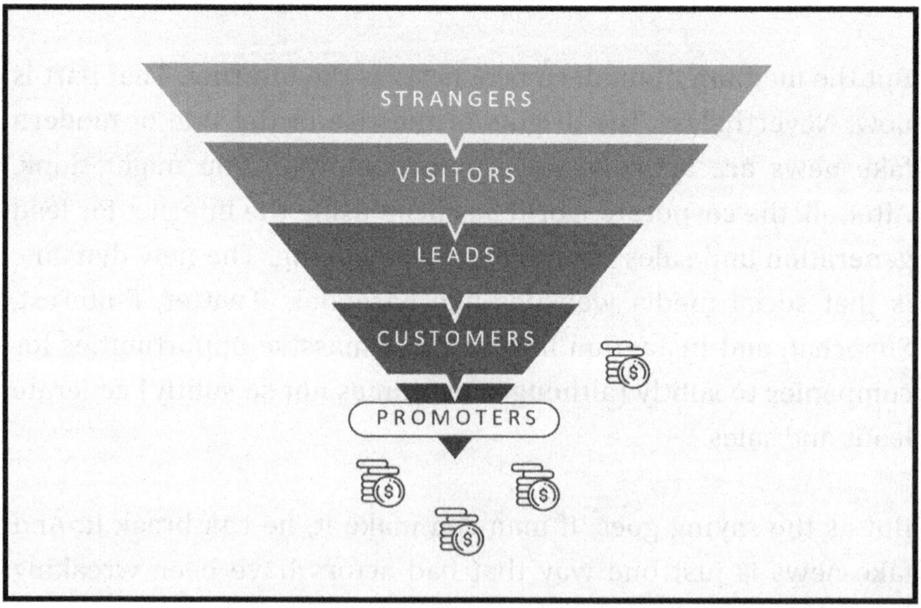

The leads attracted through content marketing enter a funnel in a way that strangers become visitors, visitors become leads, leads become clients, and clients become promoters of your brand and company, as illustrated in Figure 20-1. During this sales funnel process, it does not matter how many leads are lost. All that matters is that some people become customers — and, more importantly, that some become promoters of your brand.

One of the ways that content marketing works is by buying influence online. This is normal for a lot of businesses. They pay for advertising. They buy followers. They pay for influencers on various social media platforms to use, wear, and promote their brands. In fact, there is an entire industry built around influencer marketing. And there have been some recent issues with influencer marketing following the Fyre Festival scandal.

Figure 20-2: Influencer Marketing Price Per Post[3]

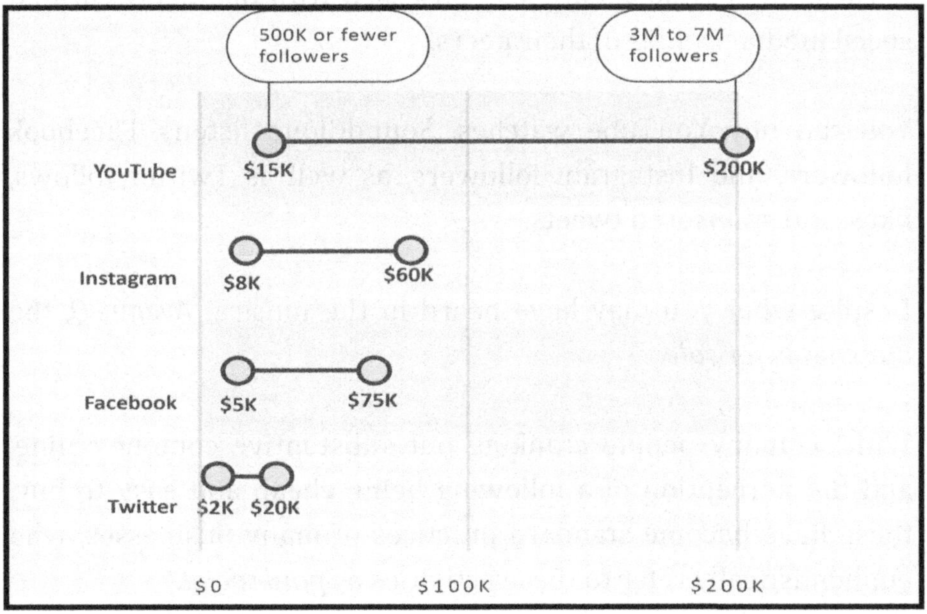

You can see the different rates of social media influencer activities in Figure 20-2. And neither paying influencers nor buying followers is expensive. It isn't illegal. And these are things you don't even need to go onto the dark web to do. You can just Google around for whatever you need in the social media world.

I once had a client who was depressed because his daughter was distraught that she had fewer Instagram followers than her friends. I told him that he should just buy her some followers.

He was shocked. And he looked at me with wide eyes. He asked where he could do this. I told him to just use Google to look up companies that sell followers.

Not only did he buy her some, but when I tell this story to groups of executives, many of them are pleased to learn this can be done. Because their children are also obsessed with the highly curated social media profiles of their peers.

You can buy YouTube watches, SoundCloud listens, Facebook followers, and Instagram followers, as well as Twitter follows, likes, and sponsored tweets.

Despite what you may have heard in the musical *Avenue Q*, the internet is *for sale*.

With so many people cranking out substantive content online, and the perception of a following being cheap and easy to buy, these have become standard practices of many businesses, who euphemistically refer to these activities as *paid social*.

But while this works really well for startups and established businesses, these dynamics of paid social often make it difficult to discern what is real and what is not.

Plus, the full and complete democratization of information dissemination means that there is no control on how information is spread. And so far, social media companies have been able to shirk many of their responsibilities in policing the fakeness of their platforms. To echo a notion from the previous chapter, part of the reason that Facebook and other social media platforms do not remove fake profiles as rigorously as they could is because there is significant moral hazard.

There is a conflict of interest because social media company valuations are based on the number of users, followers, and activity. So removing users, erasing fake followers, and stopping automated bot activity could negatively affect shareholder value. In other words, the fiduciary duty of executives of social media companies to their shareholders runs counter to controlling, reducing, or eliminating the dissemination of fake news and the proliferation of fake followers.

What If

While the use of social media to generate marketing and sales leads is a fairly standard process, what if it wasn't a business interest that was trying to gain attention and followers? What if a terrorist entity or a subversive political entity was trying to build its following, spread its message, and create its own kind of brand affinity?

While you can see the sales funnel for a corporate entity in Figure 20-1, there can be more insidious uses of this same effective technique, as seen in Figure 20-3. To again quote Hubspot: inbound marketing is *"the best way to turn strangers...into promoters."* This begs the question: What does the inbound and content marketing sales funnel of a terrorist entity look like? What does it look like for a politically subversive bad actor?

There are three big dynamics that make social media the equivalent of a digitally weaponized political IED:

Fake news is easy.

The internet is for sale.

Social media is cheap.

And these three dynamics threaten to make fake news an even bigger threat in the future.

Figure 20-3: Terrorist Funnel

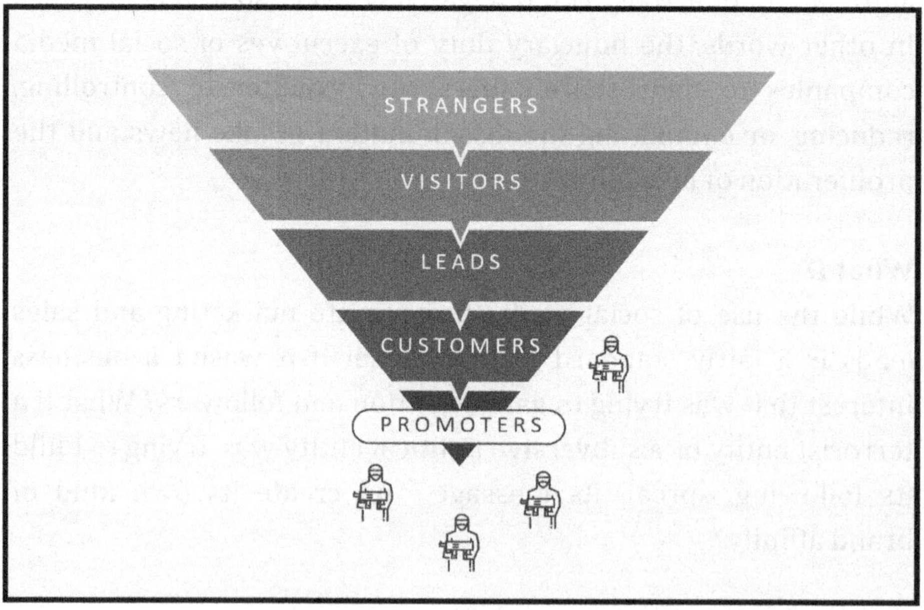

CHAPTER 21

THE FUTURE OF FAKE NEWS

The future of fake news could be very bleak indeed. The current state of political manipulation and disinformation is dire enough given the easy, cheap, and widespread ability for bad actors to abuse social media, given a major conflict of interest.

In the immediate term, elections are likely to become more socially and politically vitriolic. The 2016 U.S. presidential election and 2018 midterm elections were disgusting. This happened, in part, because the social media algorithms are designed to increase the time you are on their platforms. It's what tech people call "eyes on screens." But that doesn't mean that happy or pleasant messages are what will get you to interact.

Media folk have long lived by the saying "*If it bleeds, it leads.*" This means that the most horrific stories are the ones that will come first in the paper, in magazines, or on TV. You want to get people's attention and suck them in with horrific things. Social media is similar. If it knows that you will interact more with bad things, then you will see more things that agitate you. This is one of the ways that social media is charged with breading psychosis.

Deep Fakes

One of the greatest risks ahead for the future of fake news is the manipulation of imagery and videos. *Deep fakes* is the term used to describe the ability to create shockingly real — albeit fake — videos. These are likely to become far more prevalent and a more insidious part of the fake news landscape. Instead of people living in a post-truth era, deep fakes could allow people to believe that there is truth when there is none.

Again, *this time is never different.* During the Cold War, Stalin was notorious for doctoring early Soviet photos and removing other early Communist leaders from them. And while photo manipulation has existed for a long time, the ability to create real-looking new videos could be more effective at sowing seeds of political, social, or economic discontent. Fictitious video content is likely to be just a piece of the iceberg in terms of the fraudulent use of fake data that may lie ahead for political bad actors in the field of fake news.

What Comes Next

Countries in Europe have already started to push for more regulation of social media firms and their use of data, with the Global Data Protection Regulation — the GDPR. And a bigger European regulatory push lies ahead. Because for all of the benefits of social media, Big Tech has been left in a largely laissez-faire environment, free to do as it pleases. And with the good comes the bad. Regulation may or may not be the answer, but if the current state of social media continues undisturbed, the threat to free and democratic societies posed by social media vulnerabilities could be existential. What comes next will be critical for society as a whole.

CHAPTER 22

THE ANALYTICAL ARMS RACE

The challenge of fake news is a political one. But it is not the only analytical arms race in play on a global scale. In fact, there are two major arms races going on right now — a political arms race for analytical processing power as well as a corporate one.

On a global scale, winning the race to quantum computing is a critical prerequisite for the United States to maintain global technological supremacy. But the U.S. is at risk of losing the race to quantum computing, which can be seen in Figure 22-1. Chinese patents in the field of quantum have outpaced U.S. patents.

Plus, quantum is just one of four major scientific research programs highlighted in China's National Medium- and Long-Term Science and Technology Development Plan Outline (2006-2020). U.S. concerns about this push appeared in the USTR report that preceded the Section 301 tariffs that the United States placed on China in 2018.[1] In addition to the push for quantum computing, there is a critical push to adopt genuine artificial intelligence. And as with quantum computing, the country that gets there first will have a big advantage.

Corporate Data Wars

While countries are pushing to reach major technological breakthroughs in quantum computing and AI, the same kind of arms race is also going on between companies. And the ones that have the capital to invest and build up their AI capabilities, develop quantum, and build a culture of data are going to leave their competitors in the dust. Early adopters of such technology would likely have big first-mover advantages. But they would need to be well-capitalized. Plus, one of the biggest challenges ahead will be to get companies to embrace a culture of data.

Companies with the greatest capital will be at a distinct advantage. And this could lead to a more rapid pace of corporate consolidation in the decade ahead.

Figure 22-1: Chinese Quantum Patents Outpace the US[2]

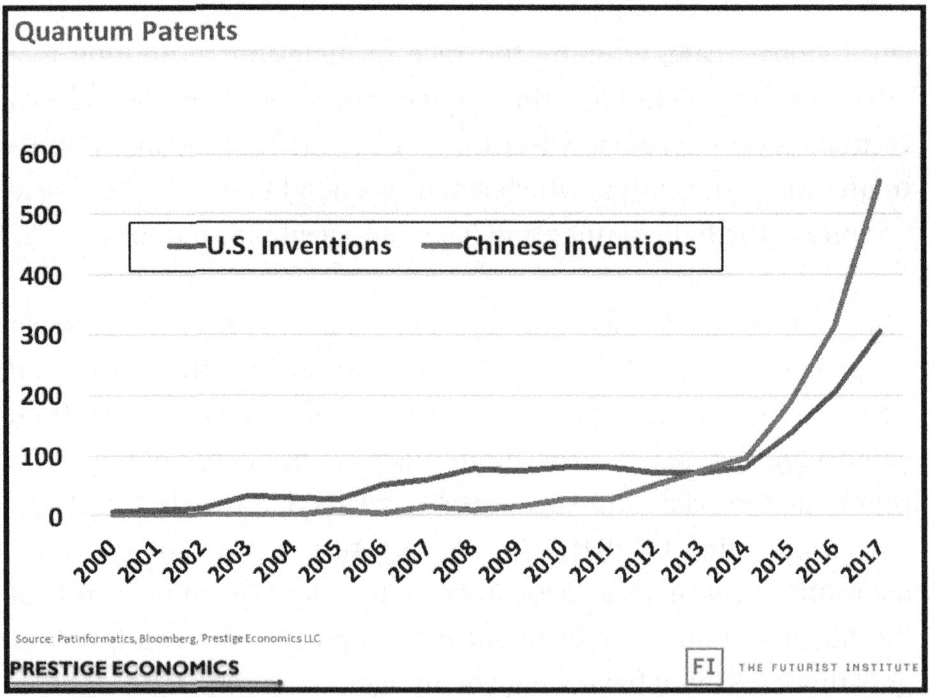

In the NewVantage Partners Big Data Executive Survey of Fortune 1000 executives from 2017, the greatest areas of realized value reported have been in decreasing expenses and finding new innovations. Some of the other most insightful data shown in Figure 22-2 reflected the least progress in using data to transform business for the future — and the least success in trying to establish a data-driven culture. In short, it appears difficult to build a culture of data.

Furthermore, while big market leaders duke it out to dominate the geopolitical or corporate world of data analytics, there is also a major risk of data poverty among less-developed nations — and among smaller companies that lack the same capital investment capabilities as the biggest firms.

Figure 22-2: Creating a Culture of Data Is Challenging[3]

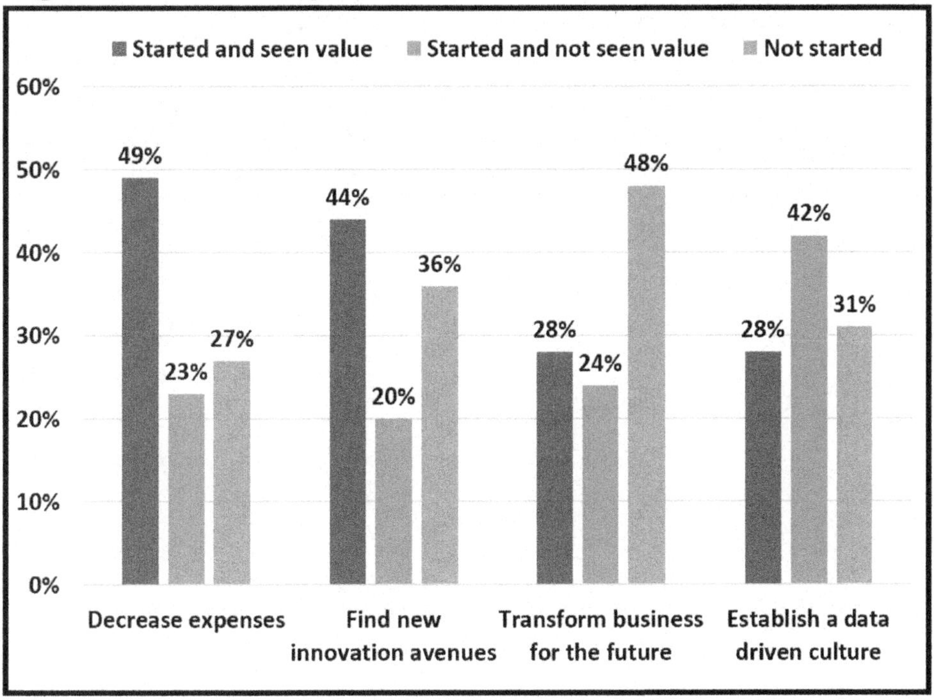

Pulling Everything Together

CONCLUSION

SEEING THROUGH THE FOG OF DATA

Thank you for reading this book on *The Fog of Data*.

I started this book with a quote from the Prussian military strategist Carl von Clausewitz that I haven't yet referenced. Now it is time.

> War is the place of uncertainty; three-fourths of the things on which the business of war rests lay in a fog of a more or less large state of uncertainty. It is therefore initially from there where a fine penetrating mind is required, with tact of judgement to feel out the truth.[1]

The reason I chose the title of this book — and the reason I've included this quote — is because working with data, for all the trappings of what is knowable and fact, is still a place of uncertainty.

More data isn't always clarifying because more data isn't always better. More, better data is better. But it's tough to know what the right data is until you start to feel around for the truth.

This is what someone working with data does. Sometimes it's more trial and error than anything else. And as I've noted in Chapters 8 and 9, sometimes you need to create your own data to find your way through the fog — to expand what is knowable and reduce that which is uncertain and unascertainable.

A related and important reality that should not be overlooked is that while some people are very excited about the current future prospects for so-called *big data*, the truth is that for many analyses, data may not yet exist. And so, the first step is actually creating data.

Somehow it seems to often get forgotten in the data world that before you get big data you need to start with *some* data — or *any* data, really. Of course, it is also often overlooked that you should consistently follow a structured data collection process to make sure that you measure twice and cut once.

You also need to keep in mind that no matter how good your analysis is, you must connect with your stakeholders. This is a *conditio sine qua non*. Use pictures and graphs — and tell stories. Don't worry if your explanations seem too simplistic. After all, you want people to understand your work, your data choices, the challenges you faced, and the actionable implications of your analysis.

Simple and easy is the way to go, if you want to make sure everyone understands you. There is nothing wrong with making things clear as crystal. Honestly, that's usually the least concern in communicating data.

And as data volumes increase, the risk that navigating that data, cleaning that data, ensuring a structures analysis process, and communicating the results could get significantly foggier.

The more moving pieces there are, the bigger the challenges will likely be in breaking things down and making them digestible. Similarly, there will be other significant data challenges ahead. And there are risks that data could be more aggressively misused. With mind-blowing volumes of data, it may be easier to hide statistical trickery, fraud, laziness, or conflicts of interest.

And dangerous fictions — in the form of fake news — could become increasingly more difficult to discern from reality. Psychosis-inducing social media, deep fakes, and influencer marketing will present significant and rising threats that go beyond data analysis. In fact, they could threaten the fabric of free and democratic societies due to an unchecked, overdemocratization of information and content generation.

Looking to the data of the future, there will be major opportunities for technological, societal, and business value creation. But there will also be increased risks.

Further Learning

If you've enjoyed this book and want to learn more about becoming a futurist and how to incorporate new and emerging technology risk into your strategic planning, I would recommend pursuing the Futurist and Long-Term Analyst (FLTA) training program that I created for The Futurist Institute. All of the details about the FLTA can be found at www.futuristinstitute.org.

ENDNOTES

Chapter 2

1. Speech by Darryl Willis, VP Oil, Gas, and Energy, Google Cloud at Google. Speech given at D2:Upheaval on 10 October 2018.
2. *Investopedia*. "Moore's Law." Retrieved on 7 November 2018 from https://www.investopedia.com/terms/m/mooreslaw.asp.
3. Gribbin, J. *Computing with Quantum Cats: From Colossus to Qubits*. Prometheus Books: New York. P. 92.
4. Jurvetson, S. (10 December 2016). "Moore's Law over 120 years." Flickr. Retrieved on 1 November 2018 from https://www.flickr.com/photos/jurvetson/31409423572/.
5.Speech by Deepu Talla, Vice President and General Manager of Autonomous Machines, NVIDIA, at RoboBusiness on 26 September 2018.

Chapter 4

1. *Hitchhiker's Guide to the Galaxy* (2005). Directed by Garth Jennings.

Chapter 12

1. Personal photo collection of Jason Schenker. Taken Philadelphia., Pennsylvania.
2. U.S. Federal Reserve System. Retrieved on 21 February 2019 from https://www.federalreserve.gov/monetarypolicy/fomccalendars.htm
3. Ibid.
4. Ibid.
5. Ibid.
6. U.S. Federal Reserve System. Retrieved on 21 February 2019 from https://www.federalreserve.gov/newsevents/speech/yellen20160826a.htm

Chapter 12

1. U.S. Department of Energy, Instructions Form EIA-803 Weekly Crude Oil Stocks Report. Retrieved on 18 February 2019 from https://www.eia.gov/survey/form/eia_803/instructions.pdf.
2. U.S. Department of Energy, Form EIA-803 Weekly Crude Oil Stocks Report. Retrieved on 18 February 2019 from https://www.eia.gov/survey/form/eia_803/proposed/form.pdf.

Chapter 14

1. Images from Adobe Stock.
2. Images from Adobe Stock.
3. Images from Adobe Stock.
4. U.S. Census Bureau, NAICS code data retrieved on 18 February 2019 from https://www.census.gov/econ/isp/sampler.php?naicscode=33392&naicslevel=5.
5. Images from Adobe Stock.
6. Images from Adobe Stock.

Chapter 15

1. Goll, H. (1876). *Die Weisen und Gelehrten des Alterthums.* Leipzig: Otto Spamer. P. 359.
2. Personal photo collection of Jason Schenker.
3. Images from Adobe Stock.
4. Personal photo collection of Jason Schenker. Taken Amarillo, Texas, July 2017.

Chapter 17

1. Image provided courtesy of Heritage Auctions, HA.com. Retrieved on 24 August 2018 from https://currency.ha.com/itm/miscellaneous/other/-1-000-000-us-treasury-bearer-bond/a/364-15547.s.

Chapter 20

1. Retrieved from www.hubspot.com.
2. Concept based on business model of Hubspot.
3. Captiv8 Data as cited in Kelly, C. (12 February 2019). "Fyre Festival to Fashion Week, How to Instagram Influencers Make So Much Money?" Retrieved from https://www.usatoday.com/story/news/investigations/2019/02/12/instagram-youtube-influencer-rates-fyre-festival-fashion-week-money-rich-branding-ads-girls/2787560002/.

Chapter 22

1. Decker, S. & and C. Yasiejko. (2018). "Forget the Trade War. China Wants to Win Computing Arms Race." Bloomberg.com, Bloomberg, 9 April 2018, www.bloomberg.com/news/articles/2018-04-08/forget-the-trade-war-china-wants-to-win-the-computing-arms-race.
2. U.S. Trade Representative Report. Pp. 10-11. Retrieved on 6 November 2018 from USTR Pp. 10-11, https://ustr.gov/sites/default/files/Section%20301%20FINAL.PDF.
3. NewVantage Partners Big Data Executive Survey (2017).

Conclusion

1. This quote, as with the German and English quotes before the Preface of this book, is sourced from Clausewitz, C. (2014). *Vom Kriege*. E-artnow. P. 44. Translated by Jason Schenker.

ABOUT THE AUTHOR

Jason Schenker is the President of Prestige Economics and the world's top-ranked financial market futurist. Bloomberg News has ranked Mr. Schenker one of the most accurate forecasters in the world in 43 different categories since 2011, including #1 in the world in 25 categories for his forecasts of the Euro, the Pound, the Swiss Franc, the Chinese RMB, crude oil prices, natural gas prices, gold prices, industrial metals prices, agricultural prices, U.S. non-farm payrolls, and U.S. new home sales.

Mr. Schenker has written 13 books and edited two almanacs. Five of his books have been #1 Best Sellers on Amazon, including *Commodity Prices 101, Recession-Proof, Electing Recession*, *Quantum: Computing Nouveau,* and *Jobs for Robots*. He also edited the #1 Best Seller *The Robot and Automation Almanac - 2018* as well as the 2019 edition of the almanac. Mr. Schenker is also a columnist for *Bloomberg Opinion.* He has appeared as a guest and guest host on Bloomberg Television as well as a guest on CNBC. He is frequently quoted in the press, including *The Wall Street Journal, The New York Times*, and *The Financial Times.*

Prior to founding Prestige Economics, Mr. Schenker worked for McKinsey & Company as a Risk Specialist, where he directed trading and risk initiatives on six continents. Before joining McKinsey, Mr. Schenker worked for Wachovia as an Economist.

Mr. Schenker holds a Master's in Applied Economics from UNC Greensboro, a Master's in Negotiation from CSU Dominguez Hills, a Master's in German from UNC Chapel Hill, and a Bachelor's with distinction in History and German from The University of Virginia. He also holds a certificate in FinTech from MIT, an executive certificate in Supply Chain Management from MIT, a graduate certificate in Professional Development from UNC, a certificate in Negotiation from Harvard Law School, and a certificate in Cybersecurity from Carnegie Mellon University.

Mr. Schenker holds the professional designations ERP® (Energy Risk Professional), CMT® (Chartered Market Technician), CVA® (Certified Valuation Analyst), CFP® (Certified Financial Planner), and FLTA™ (Certified Futurist and Long-Term Analyst). Mr. Schenker is also an instructor for LinkedIn Learning. His courses include Financial Risk Management, Recession-Proof Strategies, Audit and Due Diligence, and a weekly Economic Indicator Series.

Mr. Schenker is a member of the Texas Business Leadership Council, the only CEO-based public policy research organization in Texas, with a limited membership of 100 CEOs and Presidents. He is also a member of the 2018 Director class of the Texas Lyceum, a non-partisan, nonprofit that fosters business and policy dialogue on important U.S. and Texas issues. He is also on the Texas Lyceum Executive Committee as the VP of Technology.

Mr. Schenker is an active executive in FinTech. He has been a member of the Central Texas Angel Network and is currently the Executive Director of the Texas Blockchain Association. He is also a member of the National Association of Corporate Directors as well as an NACD Board Governance Fellow.

In October 2016, Mr. Schenker founded The Futurist Institute to help analysts, strategists, and economists become futurists through a training and certification program. Participants can earn the FLTA — Certified Futurist and Long-Term Analyst designation.

Mr. Schenker was ranked one of the top 100 most influential financial advisors in the world by Investopedia in June 2018.

For more information about Jason Schenker:
www.jasonschenker.com

For more information about The Futurist Institute:
www.futuristinstitute.org

For more information about Prestige Economics:
www.prestigeeconomics.com

TOP FORECASTER ACCURACY RANKINGS

Prestige Economics has been recognized as the most accurate independent commodity and financial market research firm in the world. As the only forecaster for Prestige Economics, Jason Schenker is very proud that Bloomberg News has ranked him a top forecaster in 43 different categories since 2011, including #1 in the world in 25 different forecast categories.

Mr. Schenker has been top ranked as a forecaster of economic indicators, energy prices, metals prices, agricultural prices, and foreign exchange rates.

ECONOMIC TOP RANKINGS

#1 Non-Farm Payroll Forecaster in the World
#1 New Home Sales Forecaster in the World
#2 U.S. Unemployment Rate Forecaster in the World
#3 Durable Goods Orders Forecaster in the World
#6 Consumer Confidence Forecaster in the World
#7 ISM Manufacturing Index Forecaster in the World
#7 U.S. Housing Start Forecaster in the World

ENERGY PRICE TOP RANKINGS

#1 WTI Crude Oil Price Forecaster in the World

#1 Brent Crude Oil Price Forecaster in the World

#1 Henry Hub Natural Gas Price Forecaster in the World

METALS PRICE TOP RANKINGS

#1 Gold Price Forecaster in the World

#1 Platinum Price Forecaster in the World

#1 Palladium Price Forecaster in the World

#1 Industrial Metals Price Forecaster in the World

#1 Copper Price Forecaster in the World

#1 Aluminum Price Forecaster in the World

#1 Nickel Price Forecaster in the World

#1 Tin Price Forecaster in the World

#1 Zinc Price Forecaster in the World

#2 Precious Metals Price Forecaster in the World

#2 Silver Price Forecaster in the World

#2 Lead Price Forecaster in the World

#2 Iron Ore Forecaster in the World

AGRICULTURAL PRICE TOP RANKINGS

#1 Coffee Price Forecaster in the World

#1 Cotton Price Forecaster in the World

#1 Sugar Price Forecaster in the World

#1 Soybean Price Forecaster in the World

FOREIGN EXCHANGE TOP RANKINGS

#1 Euro Forecaster in the World

#1 British Pound Forecaster in the World

#1 Swiss Franc Forecaster in the World

#1 Chinese RMB Forecaster in the World

#1 Russian Ruble Forecaster in the World

#1 Brazilian Real Forecaster in the World

#2 Turkish Lira Forecaster in the World

#3 Major Currency Forecaster in the World

#3 Canadian Dollar Forecaster in the World

#4 Japanese Yen Forecaster in the World

#5 Australian Dollar Forecaster in the World

#7 Mexican Peso Forecaster in the World

#1 EURCHF Forecaster in the World

#2 EURJPY Forecaster in the World

#2 EURGBP Forecaster in the World

#2 EURRUB Forecaster in the World

ABOUT THE PUBLISHER

Prestige Professional Publishing LLC was founded in 2011 to produce insightful and timely professional reference books. We are registered with the Library of Congress.

Published Titles

Be the Shredder, Not the Shred

Commodity Prices 101

Electing Recession

Financial Risk Management Fundamentals

Futureproof Supply Chain

A Gentle Introduction to Audit and Due Diligence

Jobs for Robots

Midterm Economics

The Promise of Blockchain

Quantum: Computing Nouveau

Robot-Proof Yourself

The Fog of Data

The Robot and Automation Almanac — 2018

The Robot and Automation Almanac — 2019

Future Titles

Reading the Economic Tea Leaves

Spikes: Growth Hacking Leadership

— THE ROBOT AND AUTOMATION ALMANAC —

The Robot and Automation Almanac: 2019 is a collection of essays by robot and automation experts, executives, and investors on the big trends to watch for in automation, AI, and robotics in 2019. *The Robot and Automation Almanac: 2019* was compiled by The Futurist Institute and published by Prestige Professional Publishing in December 2018.

JOBS FOR ROBOTS

Jobs for Robots provides an in-depth look at the future of automation and robots, with a focus on the opportunities as well as the risks ahead. Job creation in coming years will be extremely strong for the kind of workers that do not require payroll taxes, health care, or vacation: robots. *Jobs for Robots* was published in February 2017. This book has been a #1 Best Seller on Amazon.

ROBOT-PROOF YOURSELF

Robot-Proof Yourself offers a number of practical professional recommendations for how to be robot-proof in the coming era of professional, economic, and financial disruptions. Robots and automation are set to advance, but individuals have a chance to benefit from the coming changes. *Robot-Proof Yourself* was released in December 2017.

DISCLAIMER

FROM THE AUTHOR

The following disclaimer applies to any content in this book:

This book is commentary intended for general information use only and is not investment advice. Jason Schenker does not make recommendations on any specific or general investments, investment types, asset classes, non-regulated markets, specific equities, bonds, or other investment vehicles. Jason Schenker does not guarantee the completeness or accuracy of analyses and statements in this book, nor does Jason Schenker assume any liability for any losses that may result from the reliance by any person or entity on this information. Opinions, forecasts, and information are subject to change without notice. This book does not represent a solicitation or offer of financial or advisory services or products; this book is only market commentary intended and written for general information use only. This book does not constitute investment advice. All links were correct and active at the time this book was published.

DISCLAIMER

FROM THE PUBLISHER

The following disclaimer applies to any content in this book:

This book is commentary intended for general information use only and is not investment advice. Prestige Professional Publishing LLC does not make recommendations on any specific or general investments, investment types, asset classes, non-regulated markets, specific equities, bonds, or other investment vehicles. Prestige Professional Publishing LLC does not guarantee the completeness or accuracy of analyses and statements in this book, nor does Prestige Professional Publishing LLC assume any liability for any losses that may result from the reliance by any person or entity on this information. Opinions, forecasts, and information are subject to change without notice. This book does not represent a solicitation or offer of financial or advisory services or products; this book is only market commentary intended and written for general information use only. This book does not constitute investment advice. All links were correct and active at the time this book was published.

[illegible]

[illegible]

Prestige Professional Publishing LLC

7101 Fig Vine Cove

Austin, Texas 78750

www.prestigeprofessionalpublishing.com

PRESTIGE
PROFESSIONAL PUBLISHING

ISBN: 978-1-946197-27-6 *Paperback*
978-1-946197-21-4 *Ebook*

Made in United States
Orlando, FL
04 November 2026

90202244R00105